PHOTOVOLTAIC DEMONSTRATION PROJECTS

Proceedings of the second contractors' meeting organised by the Commission of the European Communities, Directorate-General for Energy, held in Brussels, Belgium, 28–29 April 1987

PHOTOVOLTAIC DEMONSTRATION PROJECTS

Edited by

W. B. GILLET, J. E. BATES

Sir William Halcrow & Partners, Burderop Park, Swindon, UK

and

W. KAUT

Commission of the European Communities, Directorate-General for Energy, Brussels, Belgium

ELSEVIER APPLIED SCIENCE
LONDON and NEW YORK

ELSEVIER APPLIED SCIENCE PUBLISHERS LTD
Crown House, Linton Road, Barking, Essex IG11 8JU, England

Sole Distributor in the USA and Canada
ELSEVIER SCIENCE PUBLISHING CO., INC.
52 Vanderbilt Avenue, New York, NY 10017, USA

WITH 29 TABLES AND 93 ILLUSTRATIONS

British Library Cataloguing in Publication Data
Photovoltaic demonstration projects.
1. Europe. Solar energy. Photovoltaic conversion. Demonstration projects
I. Gillet, W. B. II. Bates, J. E. III. Kaut, W.
621.31′244

ISBN 1-85166-190-5

Library of Congress CIP data applied for

Publication arrangements by Commission of the European Communities, Directorate-General Telecommunications, Information Industries and Innovation, Luxembourg

EUR 11245

Printed in Great Britain by Galliard (Printers) Ltd, Great Yarmouth

PREFACE

The Demonstration Programme of the Commission of the European Communities in the energy field started in 1978. The first five-year programme, running from 1978 to 1982, was followed by a three-year programme from 1983 to 1985. In 1986 a new four-year programme started. The legal basis for this new programme is the Council Regulation (EEC) No 3640/85 of 20 December 1985 on the promotion, by financial support, of demonstration projects and industrial pilot projects in the energy field, published in the Official Journal of the European Communities, No L350 of 27 December 1985.

This demonstration programme covers energy saving, renewable energies such as geothermal, wind and solar energy, biomass, hydro-electric power, substitution of hydrocarbons as well as liquefaction and gasification of solid fuels.

The solar sector comprises the thermal use of solar energy and photovoltaics.

This book provides an overview of the photovoltaic demonstration projects which have been supported by the Energy Directorate of the Commission of the European Communities since 1983. It includes reports by each of the contractors who submitted proposals in 1983 and 1984, describing progress with their projects. Summaries of the discussions held at the meeting, which included contractors whose projects were submitted in 1985, are also presented.

Contractors meetings provide a valuable opportunity for those involved in the realisation of photovoltaic demonstration projects to share their experiences, and to discuss the practical difficulties encountered in the field.

The different technologies which are being demonstrated concern the modules, the cabling of the array, structure design, storage strategy and power conditioning. The various applications include powering of houses, villages, recreation centres, water desalination, radio transmitters, dairy farms, water pumping, and warning systems.

The information contained in this book is therefore very much up to date and of high value to all those interested in the future of solar electricity.

H. Eliasmöller

CONTENTS

SESSION I - GENERAL

Welcome and Overview of Session I

Overview of the PV Demonstration Programme

Sesame Data Base

Monitoring PV plants

Model calculation for the economic analysis of PV demonstration projects

SESSION 1 : GENERAL

1.1 WELCOME

Mr Eliasmöller from the CEC DG XVII welcomed the participants to the second photovoltaic contractors meeting to be held by DG XVII. He hoped that the meeting would provide both an opportunity for contractors to inform the Commission on progress with their demonstration projects, and a chance for contractors to exchange information about their projects with each other.

The current four year demonstration programme, with a budget of 360 million ECU began in 1986 and will finish in December 1989.
Mr Eliasmöller reported that the 1987 call for proposals which finished at the end of April, had encouraged the submission of several proposals for photovoltaic projects, and he anticipated that the next call for tenders which would be issued at the beginning of 1988 would include a call for photovoltaic projects. An evaluation of the whole demonstration programme is currently underway. The results of this evaluation will guide the commission in its planning of a follow-on programme.

Mr Eliasmöller regretted that Mr Kaut from DG XVII was unable to attend the meeting due to illness, and introduced Mr Joels, also from DG XVII who would be taking the place of Mr Kaut during the meeting.

1.2 OVERVIEW OF SESSION 1

1.2.1 Overview of the Demonstration Programme

In the absence of Mr Kaut, Mr Joels presented an overview of the DG XVII photovoltaic demonstration programme. The presentation was based on a paper presented by Mr Kaut to a meeting at the European PV Plant Monitoring Group, in Lugano, Switzerland, in April 1987.

Mr Joels noted that 57 photovoltaic projects had been selected from the annual calls for tenders in the period 1983-86. The call for tenders in 1987 has just drawn to a close and several proposals for PV demonstration projects were expected. The PV demonstration programme covers several applications, including:

- Electrification of rural housing
- Warning systems
- Telecommunications systems
- Lighthouses
- Street Lighting
- Water Pumping and Desalination
- Electrification of remote sites

Several of the demonstration projects have experienced problems, most of which were not due to the PV generator itself. For example, installation problems at remote sites due to restricted or difficult access. Most of the projects which began in 1983 and 1984 have now solved these problems and it is hoped that by exchanging information with those contractors who are commencing projects now, such problems may be avoided in the future. An important opportunity for this exchange of information is provided by contractors meetings.

One of the main aims of the PV demonstration programme should be to show that photovoltaics is a reliable power source, and that reliable and efficient balance of system components are available for use with PV generators.

1.2.2 SESAME Data Base

Mr Joels explained that the aim of the SESAME data base was to encourage the commercialisation of the technologies which are being demonstrated. The data base which is updated regularly by the Commission, contains technical, operational and financial information concerning ongoing and completed demonstration projects. SESAME will soon be open to the public and direct access to it will be available through the international packet switching system and through services such as Videotex, Minitel Teletel and Prestel. Indirect access to the system will be available in member states through trade associations and through the information offices in government energy departments.

1.2.3 Monitoring of PV Demonstration Projects

Dr Blaesser from the CEC Joint Research Centre (JRC) in Ispra, gave an introduction to techniques for monitoring PV systems. These are specified in detail, in "Guidelines for Photovoltaic System Monitoring", available from the JRC. These guidelines have been discussed and approved by the European PV Plant Monitoring Group, which contains photovoltaics experts from each European Country. Two levels of monitoring are recommended for PV systems. For systems under 5 kWp, a simple global monitoring scheme, using for example electro-mechanical counters to monitor the energy output of the array, is sufficient. For systems with a power output of greater than 5 kWp, full analytic monitoring should be carried out. Although analytical monitoring systems are now becoming more reliable, it was recommended that all PV demonstration projects should also be fitted with simple global monitoring equipment.

Monitoring is important, not only to demonstrate that PV systems function reliably, but also to permit comparison of predicted and actual performance, and hence the validation of design methods and performance prediction models.

1.2.4 Model Calculation for the Economic Assessment of PV Systems

At the first contractors meeting of the PV demonstration programme, it was suggested that a standard method to calculate the cost of PV generated electricity should developed. The aim of the calculation should be to help show that PV technology is commercially viable. Dr Gillett outlined a model calculation which had been developed with the help of the European PV Plant Monitoring Group.

Demonstration projects may have exceptionally high costs for several reasons including to prototype manufacturing costs of components and the extra design and project management effort required. The model calculation therefore considers not only the costs of the demonstration project itself, but also the predicted costs of 10 more systems built in 1987 and in 1990. The calculation considers only the PV generator, ie PV modules, power conditioning equipment and batteries. End use equipment is not included.

OVERVIEW OF THE EUROPEAN COMMUNITY'S PHOTOVOLTAIC DEMONSTRATION PROGRAMME

W. KAUT

Commission of the European Communities
Directorate General for Energy

1. INTRODUCTION

PV Projects are one of the two subgroups of demonstration projects of the sub-programme 'Solar energy' of the Commission's overall demonstration programme in the energy field. Although this overall programme started in 1978, really valuable PV proposals have come in only since 1983.

The majority of the selected projects have involved electrification of isolated sites, an application for which PV is most attractive today. The applications vary from electrification of isolated dwellings, farm houses, mountain refuges and buildings in nature reserves, to lighthouses, telecommunications equipment, desalination plants and warning devices. Only three projects are working with real grid connection and interaction.

Some statistical data relating to the projects which were approved in the period 1983 to 1986 is given in Table 1. A summary table listing each of the projects in the programme is presented in Annex I.

	Year	1983	1984	1985	1986	1983/86
Proposals Received	Total solar (therm. + PV)	124	98	77	80	379
	PV alone	29	34	27	27	117
	PV % of total	24	35	35	33	31
Proposals Selected	Total solar (therm. + PV)	37	39	38	31	145
	PV alone	10	16	18	13	57
	PV% of total	27	41	47	42	39
	Inst. capacity in KWP	168.7	246	197	313	924.7
Budget	Total solar (therm. + PV) MioECU	8.00	5.57	6.9	6.85	27.32
	PV alone MioECU	2.20	2.98	3.3	3.75	12.23
	PV % of total	27.5	53.5	48	54.7	44.8

TABLE 1 : Solar Demonstration Projects

2. THE PV PROJECTS

2.1 Background

Photovoltaics (PV) played a minor, and relatively unimportant role in the early years of the demonstration programme. Until 1982 only 4 PV-demonstration projects had been carried out, 3 of which were successfully completed, the fourth ended as a failure. The 3 successful projects were:

- a PV-driven pump (900 Wp) for a micro-irrigation plant in the south-west of France (Final report: EUR 8571),

- an electric car, the batteries of which were charged by a PV-array of 1.8 kWp (Final report: EUR 9039),

- the interconnection of a 5 kWp PV panel with the electric low-voltage utility grid (Final report: EUR 10.187).

A remarkable increase of PV-projects among the overall solar proposals was noted in 1983 and this was confirmed in 1984, 1985 and 1986.

PV-applications are generally classified in two major groups:

- stand alone systems, without any grid connection
- grid connected systems

2.2 Stand-alone installations

The major part of all PV-demonstration projects are stand-alone installations. Most of these projects concern PV electricity supply to all kinds of houses or buildings being used:

- as dwellings
- for agricultural purposes
- for hotel and restaurant purposes
- as high-mountain refuges.

A PV system has to be adapted to the individual needs of the houses to be supplied, according to their consumption and their load profile. The size of the PV plant and the battery storage capacity must be optimised. In case of groups of houses or small villages it must be decided, if a central PV-plant with a central storage facility and distribution grid should be provided or if each house should get its own individual PV system. Will the modules be fixed on the house, in which case the overall architecture of the building should be considered, or will the modules be installed on the ground. In all cases, environmental aspects must be taken into account.

As electricity is an absolutely new factor in the inhabitants's life, an extensive information and education campaign is required in order to encourage a rational use of the precious PV electricity.

Furthermore only low-energy consumption appliances can normally be justified for PV powered buildings.

There are two major advantages offered by the decentralized PV system for small villages:

- PV electricity is not available in unlimited quantities. In case of centralized systems a certain discipline and community spirit is required of all inhabitants, when consuming energy. In case of decentralized systems each house owner has it in his own hands how to use the energy that is at his disposal. In this way tensions and mistrust among the population can be avoided.

- the distribution grid, consisting of masts, wires and cables in the air, which would in many cases cause a nasty pollution of the picturesque, natúral environment of the isolated sites, can be avoided.

Some agricultural applications are also included in the demonstration programme. A number of dairy farms of different size, located in the Italian Alps, will be equipped with PV systems. The access to these farms is extremely difficult and the milk is transported down to the valley only once per week (and this in some cases by helicopter). Meanwhile the milk must be cooled or processed on the farms.

A very interesting application of PV in isolated houses is the equipment of high-mountain refuges. Situated at very high altitudes up to 3,000m, sometimes placed like bird-nests in the rocks, most of them can only be reached by hikers or by helicopters for provisioning. Energy supply by grid connection or the use of diesel generators must be excluded both for economic and for environmental reasons. Consequently PV finds here an ideal application. The installed capacities are between 200 and 640 Wp according to the size and capacity of these refuges; the battery storage capacity is between 3,000 and 16,000 kWh. All systems are DC systems.

A number of Greek and Italian lighthouses in the Mediterranean sea will get PV equipment. Some of them are operated now with gas-fired lamps and mechanical turning devices. In this case with PV a completely new (electrical) lamp system must be installed. In other cases the existing diesel generators will be replaced by PV modules and, the diesels in future as a stand-by power source.

The main reasons for transforming these lighthouses to PV are:

- high reliability of PV:
- low maintenance needs of PV
- no or low fuel provision needs.

Many telecommunication relay-stations and TV-FM transmitters are located on the top of mountains in very isolated areas, accessible only under extreme difficulties. In the past, when no grid-connection was available, the only solution for power supply was the diesel-generator set with its well known inconveniences. Fuel provision is very difficult and hence costly under the given circumstances. A diesel requires high maintenance and hence operating personnel permanently on site. Furthermore it is not easy to find people who like to work in such extreme isolation. All these problems are similar to those encountered with the lighthouses.

Archaeological sites are often far away from any energy supply. Nevertheless energy can be needed for different purposes. The use of a fossil energy source mostly is to be rejected for environmental reasons. This offers an opportunity for PV generators. In Tuscany in Italy, two Etruscan tombs will be accessible to visitors thanks to lighting and draining systems powered by PV electricity.

Another interesting field of PV-application are warning and measuring systems, often located in isolated areas. Three projects in this sector concern fire, high-water and fog-warning and the water-level measurement in deep-wells of a water-works. The role of PV in all these cases is the energy supply for the sensors and the equipment for handling the data and also for the tele-transmission of the data to data collecting centres or central control facilities.

2.3 Grid-connected projects

Although grid connected applications are still rather uneconomical, it is felt that some experience in this field is needed.

As the electricity produced by the PV generator will be fed into the grid under certain operating conditions, this electricity must be of high quality in frequency and waveform. This in return demands rather sophisticated power-conditioning equipment.

In the framework of an Irish project, a 5.2 kWp PV generator is mounted on a one family house. The system consisting of the PV-modules, an inverter with MPPC and a transformer, is designed for full grid interaction. This generator is expected to supply under local climatologic conditions, about 5,500 kWh/year, more or less the annual consumption of a one-family household. In Germany, an 8 kWp PV generator will be installed on a house, operating with a new high efficiency inverter. In the UK a combination of PV and wind generators will be installed to supply a cluster of low energy houses.

3. MONITORING

The monitoring phase is of great importance for a demonstration project. Only through careful monitoring one can find out how well a plant is working. But the best solar plant cannot give good monitoring results if the monitoring system is wrong or inadequate in its concept or if the monitoring campaign is carried out without care or competence.

For this reason, in very close collaboration with the specialists of the CEC Joint Research Centre of Ispra, maximum advantage is taken of the experience and the know-how they acquired in the framework of the CEC R & D programme and especially of the PV-pilot programme.

'Guidelines for PV system monitoring' (Issue 3) have been elaborated which should help our contractors to design their monitoring system and to carry out the monitoring campaign in an appropriate way.

Information is given on which parameters should be measured, on the required accuracy of the instruments to be used, on the frequency of measurements, the recording format and on data handling at the JRC at Ispra.

Already in the publication of the invitation for proposals for demonstration projects, people are informed that a 2-year monitoring programme must be provided for projects proposed and that this programme must be in conformity with the above mentioned guidelines. Later on the contractors are obliged to transmit the monitoring data recorded in the appropriate manner, to Ispra regularly (in general every 1-2 months). The JRC carry out an evaluation and analysis of these data and prepares monthly summary reports for all plants.

Furthermore comparative long-duration tests are being carried out in the JRC test field on all module types used in the demonstration projects.

Finally the Ispra experts are at the disposal of our contractors in case of problems or questions related to the monitoring of their plants.

ANNEX I

Projects included in the PV Demonstration Programme

Proj Nr SE/	Country	Application	Location	Array kWp
123	IT	Fire prevention and high water alarm	L'Aguila & Pescara	0.4
134	DE	PV system in connection with small power heat cogeneration for an isolated farmhouse	Sachsenheim Baden-Würtemberg	5
215	HE	PV pumping for irrigation	Menetes Carpathos Island	10
313	HE	PV station for two remote villages	Gavdos Island	20
314	HE	PV station for groups of isolated houses	Antikythira Island	35
404	HE	PV station for a remote village	Arki Island	25
466	FR	PV rural electrification of 40 isolated houses	Southern France	32
560	DE	PV electricity supply for ground water level measurement	Hamburg	0.3
704	HE	PV electricity supply for an isolated relay station	Antikythira Island	20
9 Projects			TOTAL	147.7

EC PHOTOVOLTAIC DEMONSTRATION PROJECTS IN 1983

Proj Nr SE/	Country	Application	Location	Array kWp
084	DE/FR	PV electricity for three stand alone residences	Baden-Württemburg Alsace	2.6
267	DE	PV electricity for a bird and weather station on a small N. Sea Island	Scharhörn	4.1
507	UK	A PV solar/wind cogeneration system to contribute to domestic electricity supplies in mains connected houses	Milton Keynes	4.2 (PV only)
520	IR	Integration of a PV generator in mains connected residential systems	Farnanes, Cork	5
555	DK	PV solar/wind cogeneration plant for a remote community	Hasle, Bornholme Island	36
573	IT	PV and micro-hydro power study of an isolated farm	Campoligure, Ligur. Appenines	2.1
624	HE	PV electricity supply of an unmanned lighthouse	Lithari, Skyros Island	2.6
648	HE	PV and wind power supply of unmanned lighthouse	Methoni, Sapienza Pel.	2.6
674	IT	PV electricity supply of a lighthouse	Palmaiola Island	3.9
692	IT	PV electricity for Etruscan Archeological sites	Cetano & Sorano	26.1
696	FR	PV electricity supply for alpine lodgings and huts	French Alps	5.3
713	DE	Stand-alone PV and wind electricity supply for various purposes in recreational areas	Freiburg i.Br.	2.1
12 Projects			TOTAL	96.5

EC PHOTOVOLTAIC DEMONSTRATION PROJECTS IN 1984

Proj Nr SE/	Country	Application	Location	Array kWp
098	FR	PV electrification of a nature reserve	Camargue	12
114	FR	PV/diesel power supply to an isolated microwave relay	Toulon	1.92
122	FR	PV electrification of a TV & FM emitter	French-Guyana	20
138	IT	PV electrification of 30 isolated houses	Eolian Islands	9
143	HE	PV power supply for desalination refrigeration and lighting	Milos Island	31.9
146	IT	PV electrification of dairy farm in the mountains	Abruzzo	6.3
159	IT	PV powered fog detecting system for high risk highway	Verona	14.3
187	HE	Autonomous PV power supply for housing complex	Poros Island	4.15
280	DE	Grid-connected low power system	Saarbrücken	8
412	DE	PV electricity house supply for decentralized relief of mains supply	Berlin	5.1
501	FR	Solar PV and thermal energy for 29 high-mountain refuges	Alpes, Pyrenees	8.35
540	IT	PV electrification of 7 dairy farms in high mountains	Liguria	27.4
565	IT	PV plant for the energy supply of a lighthouse	Sicilia	18.2
593	FR	Photovoltaic water pumping in rural area	Corse	3.84
14 Projects			TOTAL	170.6

EC PHOTOVOLTAIC DEMONSTRATION PROJECTS IN 1985

Proj Nr SE/	Country	Application	Location	Array kWp
103	FR	PV electrification of isolated houses, TV repeater, harbour beacon, water sterilizer	Corse	9.76
159	FR	PV power supply for 3 telephone exchanges	Guadeloupe	12.4
215	PO	PV electrification of isolated rural dwellings	Portugal	35.0
233	ES	PV powered reverse osmosis plant	Isla del Moro	23.12
246	FR	PV electrification of light buoys at different latitudes	Atlantic Ocean, Mediterranean Sea, Antilles	24.0
261	ES	PV power plant	Tabarca Island	100.0
327	ES	PV rural electrification of 57 dwellings	Sierra de Segura	28.2
333	ES	PV power for a cheese cooperative	Trebujena	19.7
337	ES	PV rural electrification of 30 houses	Ivares, Oden Lleida	18.9
339	ES	PV and wind powered reverse osmosis sea water desalination plant	Fuerteventura	26.0
388	IT	PV powered airport signalling lights	Lucca - Tassignano	4.75
390	IT	Passive thermo-PV hybrid system for climatic control and electrification of isolated buildings	Palombara Sabina	3.0
493	FR	PV equipment programme of the International Activity Park	Valbonne-Sophia Antipolis	8.3
13 Projects			TOTAL	313

EC PHOTOVOLTAIC DEMONSTRATION PROJECTS IN 1986

SESAME - DATA BASE FOR R & D DEMONSTRATION AND TECHNOLOGY PROJECTS

K. JOELS

Commission of the European Communities
Directorate General for Energy

1. BACKGROUND

Following a request from the Directorates-General of Science, Research and Development and Energy, the SESAME data base was created in 1982 in response to a need for an on-going evaluation and improved dissemination of information about and results emerging from European Community funded R & D Demonstration and Technology projects. From the outset, SESAME was developed so that the information contained in it would be appropriate for external audiences, and recently Member States agreed to use SESAME as a Community Data base hosting information on National Demonstration projects.

2. PURPOSE

- Constitute a reliable information system.
- Guarantee the optimum use of the information and permit an on-going appreciation of projects.
- Obtain statistics which can be used to support project evaluation
- Enlarge the potential audience to whom information can be made available.

3. CONTENT

SESAME contains operational, technical, administrative and financial information relative to each project. This information is added to and updated as each of the principal steps of the project is completed:

1. Research and development project milestones.
2. Demonstration project construction, installation and monitoring phases.
3. Publication of project final report and close of contract file.
4. Further replication or development of the project.

The sectors covered at present are:

DEMONSTRATION

- energy saving
- renewable sources energy
- liquefaction and gasification
- substitution of hydrocarbons

R & D

- recycling of industrial and urban waste
- raw materials
- textiles
- environment
- energy saving

TECHNOLOGY

- hydrocarbon technology

Information includes:

- reference number, title, sector and location
- project objectives, description, state of advancement, and results
- details of the contractor, project leader and project financing
- details of equipment suppliers and manufacturers
- administrative details.

The current coverage, is from 1978 for R & D and Demonstration projects and from 1975 for Hydrocarbon Technology projects. More than 3,500 documents are currently included in the database and 4,000 will be included by the end of 1987.

4. UPDATING

During the life of each project, updating is carried out at least twice a year.

5. ACCESS

At present, SESAME is open to the services of the Commission, and the 12 Member State energy administrations and will shortly be accessible to the public on international packet switching networks.

Information can be obtained on the data base and its contents by writing to the Directorate General for Energy.

DG XVII/E
"SESAME"
Rue de la Loi 200
B-1049 BRUSSELS

MONITORING PV PLANTS

by G.Blaesser

Commission of the European Communities
Joint Research Centre - Ispra Establishment - ESTI
21020 ISPRA (VA) Italy

SUMMARY: This report gives an introduction to the techniques of photovoltaic system monitoring which are specified in more detail in the "GUIDELINES FOR PHOTOVOLTAIC SYSTEM MONITORING".

1. OBJECTIVES OF MONITORING

The main objectives of PV systems monitoring are:

- to demonstrate that PV plants are reliable energy systems,
- to develop design criteria for optimizing the complete PV system for a given site and application,
- to test performance predictions based on theoretical models of PV systems with actual data,
- to establish evaluation methods for comparing PV systems with other energy sources in economic terms,
- to identify operational procedures that optimize plant efficiency.

While for many applications these requirements can be met by keeping records of system availability, energy balances etc. over periods of days or weeks ("global monitoring"), larger systems or new applications of particular nature justify a much more detailed study ("analytical monitoring").

2.1 DATA LOGGING AT THE SITE

2.1.1. CHOICE OF EQUIPMENT:

The choice of the components of a monitoring system for a PV installation should be based on the following criteria:

- reliability,
- cost effectiveness,
- flexibility,
- availability of software.

The most important factor is reliability. If a monitoring system does not provide a regular flow of correct data it is obviously not worth its installation. Unfortunately, the past experience in the USA as well as in Europe has shown that often the monitoring system was the most unreliable part of the whole PV plant.

The monitoring system should be cost-effective,i.e. it should use low-cost standard components (if they are reliable enough) instead of more expensive special parts.

For experimental installations the monitoring system should be flexible to be adapted easily to changes in system configuration.

Such flexibility is not necessary for the majority of the fixed installations; however, a flexible system can be bought in quantities and then adapted specifically to the parameters of each plant.

Systems suitable for analytic monitoring are either

- a PC equipped with data acquisition devices, which may be A/D-converter boards installed directly in the computer or separate ("front end") instruments connected to the computer via cable and standard interfaces (RS-232, IEEE-488 etc.),

or

- a data logger which is a dedicated microprocessor system integrated with its data acquisition devices and well protected against environmental stresses such as temperature, humidity and overvoltages.

The relative advantages of either choice are as follows:

	PC	DATA LOGGER
system cost (ECU):	3000 - 6000	10000 - 20000
reliability:	medium	very good
flexibility:	very high	low
software avail.:	depends on system	built-in
power consumption:	depends on system	low

If the PC solution is chosen, it should be based on a standard design using standard software (operating system, high-level languages). In this respect the IBM PC and its compatibles, running a monitoring program written in a high level language such as Turbo-Pascal under the MS-DOS Operating System seems the optimal choice for most cases. For low power consumption portable versions should be selected.

2.1.2. PROCEDURES:

The data logging procedures determine the sampling intervals and the parameters to be measured. They are specified in the "GUIDELINES FOR PHOTOVOLTAIC SYSTEM MONITORING" (Issue 3, in print).

In particular, the sampling intervals for analytical monitoring are either 1 min with averaging over 1 hr or 10 min without averaging, and the parameters are given in TABLE I in the Annex. The parameters required for global monitoring are shown in TABLE II; the recording intervals for global monitoring are not specified but should not exceed a week.

2.1.3. PROTECTION:

Overvoltages of atmospheric or operational origin have frequently been the main cause of malfunction of data acquisition systems. While most data loggers have build-in overvoltage protection, the lines connecting the sensors to PC based systems should be fitted with such devices.

Care should be taken to operate all system components only within their limits of temperature and humidity and to avoid exposure to dust of all moving parts (floppies etc.).

2.2. DATA TRANSFER TO JRC ISPRA

For automatic handling of the large amount of data provided by stations with analytical monitoring the medium and the format used for data transfer must be standardized. Data originally recorded on other media or in different formats have to be converted before sending them to JRC Ispra. The detailed specifications are given in the "GUIDELINES FOR PHOTOVOLTAIC SYSTEM MONITORING". In most cases 5.25" or 3.5" floppy disks are the most convenient and economic transfer medium and should be chosen whenever possible.

These rules do not apply to global monitoring, where records in a written form are also acceptable.

2.3. DATA STORAGE AT JRC

As safety against accidental loss the data from analytical monitoring are stored within the JRC Ispra at two different sites, i.e. at the ESTI laboratory and at the JRC Computing Centre.

2.3.1. DATA STORAGE AT ESTI:

- Installation: DEC microVAX II, running under VMS, memory 9 Mbytes, disk space 450 Mbytes
- Storage format: all data are stored as 16-bit binary integers For each station and month an independent file is stored.The files for each year are on a separate UIC (disk partition).

2.3.2. DATA STORAGE AT THE JRC COMPUTING CENTRE:

- Installation: Amdahl 5868S running under MVS; memory 64 Mbytes, disk space ca 50 Gbytes
- Storage format: EBCDIC in the same format as the original data (limited to the mandatory records). All data for one year are grouped together in a partitioned data set, the data for one month from each station being the individual members of the set.

3. EVALUATION

3.1 MONTHLY REPORTS:

For each station with analytical monitoring the JRC Ispra produces monthly reports summarizing the energy balances and the figures of merit. Star diagrams indicate the hours of reported data and those for which irradiance and array output exceeded the threshold. Graphs show the scatter of EA versus IA, the daily sum of EA for each day of the month, maximum and mean values of TA, and the cumulative frequency distribution of IA.

3.1 YEARLY REPORTS:

Yearly reports summarize the state of the monitoring data bank, and, for each station, the yearly energy balance and the figures of merit. Such reports are produced by the JRC Ispra also for the installations with global monitoring.

4. CONCLUSION

PV plant monitoring is evolving as well as the systems themselves. The "GUIDELINES" are revised and the evaluation methods are updated. In particular, the "European Working Group on PV Plant Monitoring" with experts from all EC member states is active in standardizing such procedures. It is hoped that as a result of all these efforts design and operation of PV plants and the reliability of all components will improve so much that monitoring will become unnecessary. Until then, however, it remains an essential part of the development effort in photovoltaics.

ANNEX

TABLE I
Recorded Parameters for Analytical Monitoring

Parameter	Symbol	Units
Irradiation, global (horizontal)	IG	kWh/sqm
Irradiation, array plane	IA	kWh/sqm
Ambient temperature in the shade	TA	deg.C
Array output voltage	UA	V
Array output current (total)	JA	A
Converter output current	JC	A
Current input to battery	JBI	A
Current output from battery	JBO	A
DC line voltage (Battery voltage)	UDC	V
Current to all DC loads	JLDC	A
Inverter/rectifier DC current (+/-)	JII	A
Inverter/rectifier AC energy (+/-)	EIO	kWh
Energy to all AC loads	ELAC	kWh
Energy to(+) and from (-) grid	ER	kWh
Energy from auxiliary generator (e.g. Diesel, wind turbine etc.)	EG	kWh

Notes:

1. IG should be measured with a class 1 pyranometer, IA by means of a pyranometer or a reference device; TA should be determined within +/- 2 degree C and the electrical data within +/-2% f.s.
2. The energy consumed by the control and data acquisition systems should be considered as part of the external load and not as internal losses (i.e. included in JLDC or ELAC).

TABLE II
Recorded Parameters for Global Monitoring

Parameter	Symbol	Units
Irradiation, global (horizontal)	IG	kWh/sqm
Array output current (total)	JA	A
Energy to all DC loads	ELDC	kWh
Energy to all AC loads	ELAC	kWh
Energy to(+) and from (-) grid	ER	kWh
Energy from auxiliary generator (e.g. Diesel, wind turbine etc.)	EG	kWh
Hours load connected	TAV	hrs

Notes:

1. IG should be measured with a pyranometer or a calibrated solar cell. The electrical data shall be determined to within +/-2% f.s.
2. A separate record of IG need not be made provided the Contractor demonstrates that the records from another recording station are sufficiently representative of conditions at the site.
3. ELDC may be recorded as AH if kWh recording is not appropriate.

Contract No: XVII/7032/85/CDC/402.5

MODEL CALCULATION FOR THE ECONOMIC ANALYSIS OF PV DEMONSTRATION PROJECTS

W B Gillett and J E Bates
Sir William Halcrow & Partners Ltd
Burderop Park, Swindon SN4 OQD, UK

SUMMARY

A model calculation has been developed for use with the pv demonstration projects which are carried out with financial support from the Commission of the European Communities. Its aim is to encourage consistent and comparable economic assessments of pv projects, and hence to assist with the demonstration of economically viable applications of photovoltaics.

In the model calculation, no account is taken of the costs of end use equipment. The pv system is treated only as a power generator.

In order to take into account the difference between prototype costs for a single unit, and reduced costs for full scale production, the model calculation includes cost estimates based on the assumption that 10 similar systems are being sold at the same time in 1987 and in 1990.

For ease of comparing projects, it is suggested that the model calculation should be presented using pro-forma sheets.

1. INTRODUCTION

An important objective of any demonstration project is to demonstrate that the new technology is commercially viable. This involves a demonstration of economic viability as well as of technical viability. However, at the demonstration project stage, there are many project costs which will be exceptionally high because of

(a) extra design and project management effort
(b) demonstration features such as instrumentation and monitoring
(c) developments carried out on-the-job to improve the new technology

(d) prototype component costs incurred because the components are not yet in mass production
(e) high level staff involvement.

It was agreed at the first PV Demonstration Projects Contractors Meeting in January 1986 that a model calculation should be developed for assessing the economic aspects of PV Demonstration Projects. The format sheets presented in Annex I, which have been prepared following discussions with the CEC Working Group on PV Plant Monitoring, are intended to form the first step towards the development of an agreed 'CEC Model Calculation' which can be used to determine the costs of the PV generators in the CEC demonstration projects. It is envisaged that the costs of the end use equipment and appliances will be considered separately.

2. COMPONENT COSTS

An important contribution to the cost of a PV system comes from component (equipment) costs. These may be presented both as actual project costs and as predictions for a future date (say 1990), exclusive of taxes, and in a reference currency such as ECU.

The costs of specialised PV components may be expected to fall in future years, whilst those for conventional components may not be expected to fall significantly.

It is appropriate to consider battery costs separately, because the life expectancy of batteries is usually less than for other electrical components.

3. INSTALLATION COSTS

These vary widely, especially for remote locations. Nevertheless an indication can be given of expected replication costs for projects located in similar conditions to that of the demonstration project.

4. PROJECT MANAGEMENT AND ENGINEERING

A demonstration project will need a specialised form of project management, but all projects need some project management. The costs of the time spent on project management, including design, and engineering, supervision and commissioning need to be taken into account when determining total project costs.

5. ANNUAL CAPITAL CHARGE

The annual capital charge depends on the anticipated life of the system components and on the interest rate. For the Model Calculation it is recommended that the following values be used to compute the annual capital charges from the Capital Costs:

Life of Specialised PV and system components	=	15 years
Life of Batteries	=	8 years
Interest Rate	=	8%

(A standard annual capital charge repayment table is attached as Annex II).

6. ANNUAL AVERAGE OPERATION AND MAINTENANCE (O&M) COSTS

Operating and maintenance costs for a demonstration project are usually higher than for subsequent replication projects because of the need to maintain the monitoring equipment in a demonstration project. Hence lower costs may be anticipated for replication systems. Where system prices include a warranty or an O&M contract, care should be taken to ensure that O&M costs are not included twice in the economic analysis.

7. AVERAGE ANNUAL ENERGY OUTPUT

The energy output from a PV generator depends on its location and on the extent to which the system is utilised, as well as on the efficiency of its components:

$$\text{Annual Energy Output} = \text{Solar radiation } (\text{kWh/m}^2\text{ yr}) \times \text{Array rating } (\text{kWp}) \times \text{Annual system efficiency} \times \text{Annual utilisation factor}$$

In this formula, the **annual system efficiency** is primarily a measure of the overall efficiency of the pv system components excluding the modules, since it may be defined as:

$$\text{Annual system efficiency} = \frac{\text{maximum possible annual energy output}}{\text{solar radiation x array rating}}$$

The **annual utilisation factor** is a measure of the extent to which the energy generated is actually used by the consuming loads. Systems which are required to maintain high reliability, such as telecommunication repeater stations may be expected to operate with lower utilisation factors than those where reliability is less important, such as for powering rural televisions.

8. UNIT COST OF ENERGY GENERATED

The recommended parameter to be used for comparing the economic viability of PV generators is the unit cost of electricity generated (ECU/kWh). This may be deduced from the following equation:

$$\text{Unit energy cost (ECU/kWh)} = \frac{\text{Annual capital charge + annual O\&M costs}}{\text{annual energy output}}$$

9. DISCUSSION

Predictions for future components costs are particularly difficult for PV systems, but work is continuing in an attempt to find agreed ways of predicting these costs in future. Of course the economics of PV generation are strongly dependent on PV component costs.

The uncertainties in the O&M costs may be quite high, but the total O&M cost is likely to be small compared with the capital charge for PV systems. Hence uncertainties in O&M costs will not usually have a strong influence on the Unit Energy Cost.

Uncertainty in the annual energy output lies mainly in the uncertainty concerning the system efficiency and annual utilisation.

It is anticipated that the experience gained from monitoring the CEC demonstration projects will allow better estimates to be made for each of these parameters in future, and that these will permit the economic viability of PV generators to be calculated with greater confidence.

ANNEX I

SHEET 1

MODEL CALCULATION: ECONOMIC ANALYSIS OF PV DEMONSTRATION PROJECTS

PROJECT REFERENCE No ....

PROJECT TITLE

LOCATION ....

CALCULATION PREPARED BY: Name ....

Address

Tel: .... Telex ....

DATE OF INSTALLATION ....

DATE OF CALCULATION ....

CURRENCY EXCHANGE RATE .... PER ECU

PHOTO OF INSTALLATION

SHEET 2

MODEL CALCULATION: ECONOMIC ANALYSIS OF PV DEMONSTRATION PROJECTS

1. COMPONENT COSTS

Costs should be presented in ECU for the actual demonstration project and for future dates, exclusive of taxes.

1.A SPECIALISED PV SYSTEM COMPONENT RATINGS AND COSTS

	DEMO' PROJECT COST (ECU)	REPLICATION COST (ECU) 10 systems 1987	REPLICATION COST (ECU) 10 systems 1990
(i) PV Modules			
Array Rating .... (Wp)			
Cost per Wp	....	....	....
Total Cost (ECU)	....	....	....
(ii) DC/DC Converter			
Rating .... (W)			
Total Cost	....	....	....
(iii) Maximum Power Point Tracker			
Rating .... (W)			
Total Cost	....	....	....
(iv) Charge controller			
Rating .... (W)			
Total Cost	....	....	....
(v) Control system			
Total Cost	....	....	....
(vi) DC/AC Inverter			
Rating .... (W)			
Cost per W	....	....	....
Total Cost	....	....	....
(vii) Other	....	....	....
	....	....	....
	....	....	....
TOTAL COST (ECU)	=========	===========	===========

SHEET 3

MODEL CALCULATION: ECONOMIC ANALYSIS OF PV DEMONSTRATION PROJECTS

1.B BATTERY RATINGS AND COSTS

	DEMO' PROJECT COST (ECU)	REPLICATION COST (ECU) 10 systems 1987	REPLICATION COST (ECU) 10 systems 1990
Batteries			
* Rating .... (V)			
.... (Ah)			
Cost per (kWh)	....	....	....
TOTAL COST (ECU)	=========	===========	===========

* Based on nominal voltage and Ah rating (100 hr)

1.C CONVENTIONAL COMPONENT COSTS

	DEMO' PROJECT COSTS (ECU)	REPLICATION COSTS (ECU) 10 systems 1987	REPLICATION COSTS (ECU) 10 systems 1990
Wiring	....	....	....
Switch boxes	....	....	....
Lightning protection	....	....	....
Battery housing	....	....	....
Other	....	....	....
TOTAL COST (ECU)	=========	===========	===========

SHEET 4

MODEL CALCULATION: ECONOMIC ANALYSIS OF PV DEMONSTRATION PROJECTS

2 INSTALLATION COSTS

An indication should be given of expected replication costs for projects located in similar conditions to that of the demonstration project.

	DEMO' PROJECT COSTS (ECU)	REPLICATION COSTS (ECU)	
		10 systems 1987	10 systems 1990
Transport	....	....	....
Civil works	....	....	....
Array structure	....	....	....
Electrical works	....	....	....
Commissioning	....	....	....
Other	....	....	....
TOTAL COST (ECU)	=========	===========	===========

3 PROJECT MANAGEMENT AND ENGINEERING

	DEMO' PROJECT COSTS (ECU)	REPLICATION COSTS (ECU)	
		10 systems 1987	10 systems 1990
Design and Engineering	....	....	....
Project management	....	....	....
TOTAL COST (ECU)	=========	==========	==========

SHEET 5

MODEL CALCULATION: ECONOMIC ANALYSIS OF PV DEMONSTRATION PROJECTS

4 TOTAL PROJECT CAPITAL COSTS

The capital costs may be computed from the data presented in Sections 1 to 3 above:

	DEMO' PROJECT COSTS (ECU)	REPLICATION COSTS (ECU) 10 systems 1987	10 systems 1990
Specialised pv components	....	....	....
Batteries	....	....	....
Conventional components	....	....	....
Installation	....	....	....
Project management and Engineering	....	....	....
TOTAL PROJECT CAPITAL COST (ECU)	=========	==========	==========
CAPITAL COST (Excl Batteries) (ECU)	=========	==========	==========

5 ANNUAL CAPITAL CHARGE

The annual capital charge may be computed from a standard capital repayment table for agreed values of the anticipated life of the system components and the interest rate.

Life of Specialised PV and system components	=	15 years
Life of Batteries	=	8 years
Interest Rate	=	8%

	DEMO' PROJECT CHARGE (ECU/yr)	REPLICATION CHARGE (ECU/yr) 10 systems 1987	10 systems 1990
System excluding batteries	....	....	....
Batteries	....	....	....
TOTAL ANNUAL CAPITAL CHARGE (ECU/yr)	=========	==========	==========

SHEET 6

MODEL CALCULATION: ECONOMIC ANALYSIS OF PV DEMONSTRATION PROJECTS

6 ANNUAL AVERAGE OPERATION AND MAINTENANCE (O&M) COSTS

	DEMO' PROJECT COSTS (ECU/yr)	REPLICATION COSTS without monitoring (ECU/yr)	
		10 systems 1987	10 systems 1990
Operating Costs	....	....	....
Maintenance Costs	....	....	....
TOTAL O&M COSTS (ECU/yr)	=========	==========	==========

7 AVERAGE ANNUAL ENERGY OUTPUT

The energy output from a PV generator is calculated as:

Annual Energy Output = Solar radiation (kWh/m^2 yr) X Array rating (kWp) X Annual system efficiency X Annual utilisation factor

	DEMO' PROJECT	REPLICATION	
		10 systems 1987	10 systems 1990
Solar irradiation (kWh/m^2 yr)	.....M/P*	....	....
Rating of pv array (kWp)	.....M/P*	....	....
Annual mean system efficiency	.....M/P*	....	....
Annual utilisation factor	.....M/P*	....	....
ANNUAL ENERGY OUTPUT (kWh/yr)	=========	==========	==========

(* Delete as appropriate: M = Measured or P = Predicted data)

SHEET 7

MODEL CALCULATION: ECONOMIC ANALYSIS OF PV DEMONSTRATION PROJECTS

8 UNIT COST OF ENERGY GENERATED

The unit cost of electricity generated (ECU/kWh) may be deduced from the following equation:

$$\text{Unit energy cost (ECU/kWh)} = \frac{\text{Annual capital charge + annual O\&M costs}}{\text{annual energy output}}$$

	DEMO' PROJECT	REPLICATION	
		10 systems 1987	10 systems 1990
Annual capital charge (ECU/yr)	....	....	....
Annual O&M costs (ECU/yr)	....	....	....
Annual Energy Output (kWh/yr)	....	....	....

UNIT ENERGY COST (ECU/kWh)	=======	==========	==========

ANNEX II

Interest rates

Use of the table : the number in the table is the equal annual payment required to pay off a loan of 1 unit where the annual compound interest rate paid is shown at the head of a column and the number of years for which payment is made is shown at the side of a line. For example, where the annual compound interest rate paid is 8.5 %. 1 unit is paid off after 5 years by an equal annual payment of 0.25377

Equal annual payment to pay off a loan of 1 unit   Mortgage repayments

Years	% rate 8.0	% rate 8.5	% rate 9.0	% rate 9.5	% rate 10.0	% rate 11.0	% rate 12.0	% rate 13.0	% rate 14.0	% rate 15.0	Years
1	1.08000	1.08500	1.09000	1.09500	1.10000	1.11000	1.12000	1.13000	1.14000	1.15000	1
2	0.56077	0.56462	0.56847	0.57233	0.57619	0.58393	0.59170	0.59948	0.60729	0.61512	2
3	0.38803	0.39154	0.39505	0.39858	0.40211	0.40921	0.41635	0.42352	0.43073	0.43798	3
4	0.30192	0.30529	0.30867	0.31206	0.31547	0.32233	0.32923	0.33619	0.34320	0.35027	4
5	0.25046	0.25377	0.25709	0.26044	0.26380	0.27057	0.27741	0.28431	0.29128	0.29832	5
6	0.21632	0.21961	0.22292	0.22625	0.22961	0.23638	0.24323	0.25015	0.25716	0.26424	6
7	0.19207	0.19537	0.19869	0.20204	0.20541	0.21222	0.21912	0.22611	0.23319	0.24036	7
8	0.17401	0.17733	0.18067	0.18405	0.18744	0.19432	0.20130	0.20839	0.21557	0.22285	8
9	0.16008	0.16342	0.16680	0.17020	0.17364	0.18060	0.18768	0.19487	0.20217	0.20957	9
10	0.14903	0.15241	0.15582	0.15927	0.16275	0.16980	0.17698	0.18429	0.19171	0.19925	10
11	0.14008	0.14349	0.14695	0.15044	0.15396	0.16112	0.16842	0.17584	0.18339	0.19107	11
12	0.13270	0.13615	0.13965	0.14319	0.14676	0.15403	0.16144	0.16899	0.17667	0.18448	12
13	0.12652	0.13002	0.13357	0.13715	0.14078	0.14815	0.15568	0.16335	0.17116	0.17911	13
14	0.12130	0.12484	0.12843	0.13207	0.13575	0.14323	0.15087	0.15867	0.16661	0.17469	14
15	0.11683	0.12042	0.12406	0.12774	0.13147	0.13907	0.14682	0.15474	0.16281	0.17102	15
16	0.11298	0.11661	0.12030	0.12403	0.12782	0.13552	0.14339	0.15143	0.15962	0.16795	16
17	0.10963	0.11331	0.11705	0.12083	0.12466	0.13247	0.14046	0.14861	0.15692	0.16537	17
18	0.10670	0.11043	0.11421	0.11805	0.12193	0.12984	0.13794	0.14620	0.15462	0.16319	18
19	0.10413	0.10790	0.11173	0.11561	0.11955	0.12756	0.13576	0.14413	0.15266	0.16134	19
20	0.10185	0.10567	0.10955	0.11348	0.11746	0.12558	0.13388	0.14235	0.15099	0.15976	20
21	0.09983	0.10370	0.10762	0.11159	0.11562	0.12384	0.13224	0.14081	0.14954	0.15842	21
22	0.09803	0.10194	0.10590	0.10993	0.11401	0.12231	0.13081	0.13948	0.14830	0.15727	22
23	0.09642	0.10037	0.10438	0.10845	0.11257	0.12097	0.12956	0.13832	0.14723	0.15628	23
24	0.09498	0.09897	0.10302	0.10713	0.11130	0.11979	0.12846	0.13731	0.14630	0.15543	24
25	0.09368	0.09771	0.10181	0.10596	0.11017	0.11874	0.12750	0.13643	0.14550	0.15470	25
26	0.09251	0.09658	0.10072	0.10491	0.10916	0.11781	0.12665	0.13565	0.14480	0.15407	26
27	0.09145	0.09556	0.09973	0.10397	0.10826	0.11699	0.12590	0.13498	0.14419	0.15353	27
28	0.09049	0.09464	0.09885	0.10312	0.10745	0.11626	0.12524	0.13439	0.14366	0.15306	28
29	0.08962	0.09381	0.09806	0.10236	0.10673	0.11561	0.12466	0.13387	0.14320	0.15265	29
30	0.08883	0.09305	0.09734	0.10168	0.10608	0.11502	0.12414	0.13341	0.14280	0.15230	30
31	0.08811	0.09237	0.09669	0.10106	0.10550	0.11451	0.12369	0.13301	0.14245	0.15200	31
32	0.08745	0.09174	0.09610	0.10051	0.10497	0.11404	0.12328	0.13266	0.14215	0.15173	32
33	0.08685	0.09118	0.09556	0.10000	0.10450	0.11363	0.12292	0.13234	0.14188	0.15150	33
34	0.08630	0.09066	0.09508	0.09955	0.10407	0.11326	0.12260	0.13207	0.14165	0.15131	34
35	0.08580	0.09019	0.09464	0.09914	0.10369	0.11293	0.12232	0.13183	0.14144	0.15113	35
36	0.08534	0.08976	0.09424	0.09876	0.10334	0.11263	0.12206	0.13162	0.14126	0.15099	36
37	0.08492	0.08937	0.09387	0.09843	0.10303	0.11236	0.12184	0.13143	0.14111	0.15086	37
38	0.08454	0.08901	0.09354	0.09812	0.10275	0.11213	0.12164	0.13126	0.14097	0.15074	38
39	0.08419	0.08868	0.09324	0.09784	0.10249	0.11191	0.12146	0.13112	0.14085	0.15065	39
40	0.08386	0.08838	0.09296	0.09759	0.10226	0.11172	0.12130	0.13099	0.14075	0.15056	40
41	0.08356	0.08811	0.09271	0.09736	0.10205	0.11155	0.12116	0.13087	0.14065	0.15049	41
42	0.08329	0.08786	0.09248	0.09715	0.10186	0.11139	0.12104	0.13077	0.14057	0.15042	42
43	0.08303	0.08763	0.09227	0.09696	0.10169	0.11125	0.12092	0.13068	0.14050	0.15037	43
44	0.08280	0.08741	0.09208	0.09678	0.10153	0.11113	0.12083	0.13060	0.14044	0.15032	44
45	0.08259	0.08722	0.09190	0.09663	0.10139	0.11101	0.12074	0.13053	0.14039	0.15028	45
46	0.08239	0.08704	0.09174	0.09648	0.10126	0.11091	0.12066	0.13047	0.14034	0.15024	46
47	0.08221	0.08688	0.09160	0.09635	0.10115	0.11082	0.12059	0.13042	0.14030	0.15021	47
48	0.08204	0.08673	0.09146	0.09623	0.10104	0.11074	0.12052	0.13037	0.14026	0.15018	48
49	0.08189	0.08659	0.09134	0.09613	0.10095	0.11067	0.12047	0.13033	0.14023	0.15016	49
50	0.08174	0.08646	0.09123	0.09603	0.10086	0.11060	0.12042	0.13029	0.14020	0.15014	50
51	0.08161	0.08635	0.09112	0.09594	0.10078	0.11054	0.12037	0.13026	0.14018	0.15012	51
52	0.08149	0.08624	0.09103	0.09586	0.10071	0.11049	0.12033	0.13023	0.14015	0.15010	52
53	0.08138	0.08614	0.09094	0.09578	0.10064	0.11044	0.12030	0.13020	0.14014	0.15009	53
54	0.08127	0.08605	0.09087	0.09571	0.10059	0.11039	0.12026	0.13018	0.14012	0.15008	54
55	0.08118	0.08597	0.09079	0.09565	0.10053	0.11035	0.12024	0.13016	0.14010	0.15007	55

SESSION II - HOUSING PROJECTS

SESSION 2 : HOUSING PROJECTS

2.1 OVERVIEW OF SESSION 2

There are currently 27 housing projects in the demonstration programme. These range from the electrification of mountain refuges and single rural houses to centralised PV generators for clusters of houses or small island communities. Some of the systems have no auxiliary power source, others are grid connected or have a backup diesel generator. Presentations were made on the following 1983 and 1984 projects.

(SE/134/83) Photovoltaic System Linked to a Small Power and Heat Generator

Mr Laukamp from the Fraunhofer Institute described progress with this project to provide a 2.3 kWp PV generator for a farmhouse. Unfortunately, planning permission to build a greenhouse structure on which to mount the PV modules at the proposed site had been refused, so the project was transferred to another site. Construction is now under way, and it is planned that the system should be operational in July. The power and heat co-generation part of the project has been postponed until the performance of an existing diesel generator and wood burning heating system have been established. The diesel generator will be used to supply peak electrical loads including a washing machine. The PV system uses a 5 kW inverter specially developed by the Fraunhofer Institute.

(SE/466/83) Rural Electrification of 40 Houses by PV Generators

Mr Aubert from Renault Automation reported that this programme to electrify 40 houses in rural areas of France is almost complete. Four 'types' of PV systems have been installed, a 400 Wp DC system, a 800 Wp DC system, a 800 Wp AC system and a 1200 Wp AC system. The choice between dc and ac systems was made in consultation with the users. Monitoring of all the installed systems has begun using electro-mechanical counters. The user is required to read the counter each week, and note the reading on a reporting form, which he then sends back by post.

Large variations have been observed in the electricity consumptions of different users. In some cases, the user has limited financial means and initially uses only lighting. Such users take some time to buy electrical appliances. Their consumption is therefore well below the maximum output of the PV generator and increases only slowly. Other users are easily able to buy new appliances and their electricity consumption quickly reaches the maximum amount supplied by the PV generator. For such users the PV generator may appear to have been under-designed.

It was reported that a special workshop is to be held on the electrification of rural housing. It will be held in October 1987 in Valbonne, France and will be based on the experience gained in this project.

(SE/084/84) Stand-Alone Residential PV System with Motor/Generator Backup

A 1.44 kWp PV generator has been installed at a remote youth centre in Alsace and is fully operational. Backup power is provided by a motor generator driven by butane gas. Mr Roessler from the Fraunhofer Institute reported that the data acquisition system had been damaged by over-voltages, but that over-voltage protection devices had now been fitted and monitoring had restarted. The monitoring system is powered by a PV module.

The PV generator supplies both dc and ac loads, with the 24V/220V dc/ac inverter switching into operation only when a 220V ac appliance is plugged in. The diesel generator is used when 3 phase ac is required in the centre's workshop, or when the system battery is in deep discharge.

This project originally included the electrification of two other houses. One house has had to be dropped from the project, and the other house unfortunately burned down.

(SE/507/84) Solar/Wind Co-Generation for Urban Residential Use

Mr Dichler and Mr Wolfe from Solapak reported on progress with this PV and wind cogeneration project, which will supply 9 houses in an urban setting in the UK.

The wind generator, rated at 22 kW has been installed and is operational. PV arrays have been installed on one house and on the roof of the control building The installation of PV arrays on the conservatory roofs of the other 8 houses is scheduled to be complete by mid 1987. The PV arrays will total 4.5kWp. The PV generators and wind generator are expected to act as complementary power sources, with the wind generator providing more power in the winter and the PV generator more power in the summer.

The Open University will monitor the system when it is operational. Preliminary results were reported to be available on the operation of the wind generator.

(SE/520/84) Integration of a PV Generator in an Existing Residential System

Mr McCarthy from the National Microelectronics Research Centre in Cork, described the 5.5 kWp PV generator installed in a house in Rosemount.

The PV arrays are mounted on the roof of a shed adjoining the house. The PV generator is connected to the grid through a 7 kVA line-commutated inverter, so that excess power can be fed to the grid.

The electricity utility required safety equipment to be installed between the PV generator and grid, to ensure that the PV system was automatically disconnected from the grid when pre-defined limits of voltage and frequency were exceeded. By selecting a line-commutated inverter rather than a self-commutated inverter, it was possible to incorporate the safety requirements specified by the utility into the design of the inverter. It was reported that the equipment needed to implement the requirements of the utility would otherwise have been very costly. A data monitoring system based on a microcomputer has been installed and monitoring began in late 1986.

(SE/573/84) Isolated Energy System For Power Supply of an Isolated Farm Located on Ligurian Appenines

The aim of this project was to provide electricity to an isolated farm in the Ligurian Appenines using a PV generator and a micro-hydro system. Mr Maccio from CESEN reported that the PV system was installed in September 1985 and is working well. However, the micro-hydro system had to be dropped from the project due to objections from the municipal authority on the grounds that the water source was also needed by neighbouring households. The pre-existing diesel generator is therefore used as a back-up generator and to supply the 3 phase ac needed to operate a dishwasher.

The 2.1 kWp PV arrays are installed about 50 metres from the house on 3 support structures, which provide an adjustable tilt angle. This is altered twice a year by the system users. A 3.5 kVA inverter provides single phase ac at 220V from a 120V dc input. Problems were experienced initially with the inverter, but these were solved by tropicalizing the components. An ac system was chosen partly because a high voltage was needed between the array and the inverter due to the distance of the array from the houses and partly because the houses already contained several ac appliances and a wiring system connected to the diesel generator.

(SE/713/84) Demonstration Projects for Stand-Alone Systems with Renewable Energy Sources

Mr Roessler from the Fraunhofer Institute described this project in which power is supplied to an information pavilion and water pumping system in a recreational area in Freiburg. The electrification of the information pavilion has been postponed to 1987 due to budgetary problems, but the PV pumping system has been operational since April 1986. Two water pumps are operated from batteries charged by the 1.73 kWp array. Batteries were included in the system to allow the use of currents from the array which were too low to power the pumps directly. In spring and autumn, additional power is supplied by a small wind generator.

A special start/stop control system had to be developed for the pumps which are designed to be connected directly to a PV array. The control system prevents the pumps from drawing too large a current from the battery and protects the pumps from a drop in water levels and any fault in the systems electronics. The control system proved more difficult to develop then was expected.

Monitoring of the system began in October 1986.

(SE/696/84) Photovoltaic Equipment for 12 Lodgings and Refuges Located in the French Alps

Mr Salecki from APREN reported that all 12 PV systems had been installed in mountain refuges. 3 sizes of system were used in the project, 280 Wp, 480 Wp, and 600 Wp. The orientation of each array was decided in consultation with the refuge warden and at two sites, some modules face east to take advantage of a clear morning sky, which often clouds over later in the day. Results are not yet available to confirm the extent to which this has increased the annual outputs of these two systems.

At refuges which experience particularly low temperatures, the system batteries are enclosed in a thermally insulated box to protect them. A small electric resistance heater, controlled by a thermostat maintains the battery temperature at a suitable level.

10 of the systems are monitored using watt hour meters, which the refuge warden reads each day. The other two systems are being monitored more fully. Initial results from the monitoring show that in general the electricity consumption is much lower than estimated. Most wardens are only using lights and a small TV, whereas the systems were sized to power a refrigerator as well.

(SE/313/HE) Gavdos Photovoltaic Power Plant

Mr Chadjivasiliadis from the Public Power Corporation in Greece, and Mr Kadella from Siemens Solar (previously Interatom) gave an overview of this project to supply electricity to two villages on the Greek island of Gavdos. The 20 kWp PV plant and transmission lines were installed and the control house finished in October 1986.Commissioning of the system was not possible over the winter due to the weather and is scheduled for May 1987. A 30 kVA pulse width modulated dc-ac inverter, which has a high efficiency at partial loads will be used in the system.

(SE/314/83) Antikythira Photovoltaic Power Plant

A report on this project was made by Mr Chadjivasiliadis, and Mr Lowalt from AEG. A 25 kWp central PV plant will supply power for 3 small communties on the Greek Island of Antikythira. 12 individual houses will also be installed with 690 Wp stand alone PV generators which will provide a 24V dc supply. The central plant is connected to a 30kVA self commutated dc/ac inverter with a 220/380V output. The PV plant is now in operation and is providing power for street lighting.

(SE/404/83) Arki Island Photovoltaic Power Plant

A 40 kWp generator will provide power for 38 consumers, including a church and shops, in the small port on Arki Island. Mr Chadjivasiliadis reported that construction of the plant was delayed last year, due to difficulties in finding a local contractor to undertake the civil works. These are now under way, and it is hoped that the plant will be installed by the end of August and commissioned in September1987. It is hoped that it will be possible to lower the power demand on the island by encouraging the use of high efficiency lighting.

(SE/555/84) Solar and Windpower Plant in Hasle on the Island of Bornholme

Ms Vooraad from the municipal authority of Hasle, in the Danish Island of Bornholme, described this project to provide power to Hasle from a 36 kWp PV generator and two 99 kW wind generators. The project, which is still in the planning and pre-design stage will be finalised in 1988. It was reported that a local plant for making PV modules in Denmark was under construction, and it was anticipated that 25% of the PV array fro the Hasle project would be fitted with Danish modules. The system will be monitored according to the guidelines issued by JRC Ispra.

CONTRACT NR SE/134/83

PHOTOVOLTAIC SYSTEM LINKED TO A SMALL POWER HEAT COGENERATION

G.BOPP, H.LAUKAMP, J.SCHMID
Fraunhofer Institut für
Solare Energiesysteme (ISE)
Oltmannsstr. 22
D - 7800 Freiburg

Summary

The original plan to install a photovoltaic system linked to a small power heat cogeneration system had to be cancelled. The concept has been modified and transfered to a new site. Here construction is under way. The complete system shall be operational at the end of 1987.

1. INTRODUCTION

The plans on a photovoltaic system linked to a small power heat cogeneration had to be modified. The selected site (project partner Herr Stockbrink, Alte Häfnerhaslacher Mühle, 7123 Sachsenheim 5) has been cancelled due to several reasons:

- The project partner did no longer approve the integration of the photovoltaic modules into the roof construction, a solution, which he had agreed to in former discussions.
- Alternatively was planned to construct a greenhouse, which should serve as a mounting structure for the solar-cell panels. The application for the construction of the greenhouse has been turned down by the local town administration, hence this was no solution either.
- At the owners request has the building been connected to the public grid.
- The cooperation with the project partner turned out to be rather difficult.

On the approval of the Commission of the European Community (CEC) in December 1986 the project was transfered to another site and another project partner

(Rappenecker Hof, Herr Riesterer, 7801 Oberried-Geroldstal). The "Rappenecker Hof" is an old farm in the Black Forest near the city of Freiburg.

The now selected site offers several advantages:

1. There is no public grid available. A back-up diesel generator operates for electric power supply.

2. Integration of the solar-cell panels into the roofing is easily possible.

3. The local town administration as well as the project partner have authorized the direct integration of the solar-cell panels into the roof.

4. Cooperation with the project partner runs well.

Since the whole system is still under construction, the following description will be rather brief.
The system consists of a photovoltaic generator wired in 5 strings for different voltages to allow the use of an inverter developed by our institute. These different strings are connected to corresponding batteries, which again feed the inverter powering the house grid. The photovoltaic system is backed-up by a diesel generator.

Installation of a power heat cogeneration has been postponed till accurate data on the use of the diesel engine are available. First data indicated that waste heat from the diesel engine could cover approx. 10% of the buildings heating energy demand. Thus the technical and financial effort seemed not to be worthwhile.
After one year of operation we will review the decision on power heat cogeneration and based on the gathered data and experiences we will decide, how to proceede eventually.

At present the diesel generator, the battery banks and the inverter are installed and operational. The solar generator is being installed these days and scheduled to start operation by the end of June 1987. The monitoring system shall be operational at the end of this year.

2. PHOTOVOLTAIC ARRAY

96 solar-cell modules typ AEG PQ 10/40/01 are to be employed. They will be directly mounted on the roofing structure at the south-facing side of the building. This structure is tilted at an angle of 53° to the horizontal plane.

The manufacturer reports under SOC (25°C module temperature, 1000 W/m^2 insolation) 38.4 W of output power per module, whereas our own tests yielded 34 W per module. In operation, NOC (47°C, 800 W/m^2), we found the output 24 W per module. Hence we expect a peak power of 2.3 kW from the total number of 96 modules.

3. POWER CONDITIONING

The building is wired for a 220 V AC grid, which originally had been powered by a diesel generator. Now the house grid is fed by a 5 kW inverter developed at our institut. This inverter offers a very good efficiency, which is well above 90% even at an output power of 10% of its nominal value.
The inverter's DC-busbar is fed by a battery bank with a total storage capacity of 12 kWh. It is divided into five groups of different voltage levels according to the working principle of the inverter. Each battery group will be charged by a dedicated string of solar-cell modules via a blocking diode. Charge control is performed by separate charge controllers. These controllers protect the batteries from deep discharge by starting the diesel generator and from overcharging by short-circuiting some of the solar-cell panels of a series string.

If the insolation is insufficient to cover the power demand of the house, a back-up diesel generator of 16 kW nominal power will be used, which also takes care of heavy loads and simultaneously recharges the batteries.

At present the batteries are charged solely by the diesel generator, usually while the washing machine is used. In 1985 the diesel generator had been running for about 350 hours. With the photovoltaic generator operating we expect that the operation time of the diesel engine will be reduced to about 100 hours per year determined mainly by the use of the washing machine.

4. DETAILS OF PV USER APPLIANCES

All the electrical house equipment is for 220 V/ 50 Hz operation:

- lighting with high efficiency fluorescent lamps
- kitchen appliances
- refrigerator
- deep freezer
- washing machine
- TV set
- Radio set

The load characteristics are widely unknown, since the use of the building has drastically changed from farming to a mixture of restaurant and mountain hut.
The load will be largely influenced by the number of guests, which again depends on seasons, weather conditions as well as publicity. However, we expect a favourable correlation between the number of visitors and the insolation.

5. MONITORING SCHEME

The monitoring scheme is still in the design stage, so that a detailed information cannot be given. However, it will be designed according to the ISPRA's "GUIDELINES FOR PHOTOVOLTAIC SYSTEM MONITORING" for systems over 3 KW power.
The monitoring system will be realized by using a hand-held computer HP 71 from Hewlett Packard with very low power consumption and a Multi Function Box, System Kräckmann, with a relay multiplexer, pulse counters and a digital multimeter.

The whole project had to be modified. This led to a remarkable delay in the monitoring phase. Reliable data should be available at the end of 1987 and continously gathered till the end of 1989.

6. EXPERIENCE OBTAINED DURING INSTALLATION AND COMMISSIONING

Not applicable.

7. EXPERIENCE OBTAINED DURING OPERATION

Not applicable.

CONTRACT NR SE 466/83

PROGRAMME OF RURAL ELECTRIFICATION FOR 40 HOUSES BY PHOTOVOLTAIC GENERATORS

P. COROLLER
Agence Française pour la Maîtrise de l'Energie
Avenue Emile Hugues - 06565 VALBONNE CEDEX - France

and

B. AUBERT
RENAULT Automation/Séri
2, Avenue du Vieil Etang - BP 19 - 78391 BOIS D'ARCY CEDEX - France

* * * *

SUMMARY

The project aims to study - on the significant scale of 40 installations standardized by 400 Wp, 800 Wp, 200 Wp power steps - the different aspects (sociological, economic, technical and administrative) related to electrification by photovoltaïc generators of houses in very isolated rural areas.
The various contacts with the user were occasions to inform him of the advantages and disadvantages of photovoltaïc generators and of the necessity of managing his energy needs with regard to solar irradiation. In particular, the choice between alternative or direct current was made in consultation with the user. The components (modules, lead-acid batteries, inverters, monitoring systems) have been precisely specified and tenders were invited for all the programme's requirements in this respect. The battery charge regulators have been furnished by several general contractors according to specifications. In the same way, local installers were chosen to erect the photovoltaïc generators. Today, 90 per cent of the generators are in operation, producing electricity for the users. Each week, the users have to fill out a form with data read from the amphourmeters of their data acquisition systems. Unfortunatelly, on the one hand, some counters are not in working order and on the other hand, the forms are not returned on a regular basis. It is therefore difficult to have long series of reliable data. But the real success of this project is the total satisfaction of the users.

1. INTRODUCTION

The Programme must demonstrate that power production by means of photocells can appear as an interesting alternative, both technically and economically, of the development of geographical areas where certain problems prevent the supply of power in the conventional way.

Effectively, the usual connection to the grid of isolated houses is very expensive for the community as well for the setting up of the electrical wires (more than 15 000 ECU/km), as later for the operation and the maintenance of uneconomic mains (the electric consumption of so isolated users is very low).

This project thus aims to explore, on a meaningful scale of 40 photovoltaïc generators standardized by power steps (400 Wp, 800 Wp, 1 200 Wp), the different aspects of this method of rural electrification.

The projet consisted of 5 main tasks :

- an important administrative and financial task aiming at adapting traditional ways of financing rural electrification in France to the photovoltaïc solution. These efforts have resulted in the establishment of files and the obtainment of the funds necessary for the electrification of 40 houses,
- a second task involved contacts with the users both during research into sites and during the various stages of the construction of the generator and during technical acceptance. The photovoltaïc generator had to be accepted, as completely as possible, by the user on a practical and psychological level taking into account the system's advantages and limits. AFME, its representatives, the contractors and the installers have spared no effort to this end,
- a third task was the technical coordination of the programme involving on the one hand validation of the sites and dimensioning of the installations according to general criteria (cost of the grid connection related to the remoteness of the network and to environmental constraints, social and economic value for the satisfaction of the user's energy requirements, motivation of the user). On the other hand, it aimed to obtain a standardisation of the installations which could, however, be adapted as far as possible to the needs of the users. This is why in addition to the list of components common to the whole programme, detailed specifications have been established for the other components (battery state of charge controller, for instance) and the operations necessary (mechanical and electrical assembly) for the construction of photovoltaïc generators. This has contributed to the quality of the installations,
- more than twelve photovoltaïc general contractors and installers was entrusted with the manufacture and erection of the various generators according to the rules of the programme,
- the last task involves the collection and interpretation of data. This data is also delivered to the Joint Research Laboratory of Ispra (Italy). AFME in particular has studied in detail the data provided by certain users who appear to make the most interesting use of the system. Unfortunately, these users, who live in isolated areas of the French countryside, often have limited financial means and their consumption of electricity increases only slowly. In effect, apart from several lighting devices, they cannot very quickly acquire the necessary implements (refrigerators, television sets, etc ...).

Today, 90 per cent of installations are in place and in operation and the financial and administrative formalities have been settled with regard to those remaining to be installed. Efforts are now concentrated on observing the work of the installations and maintaining them.

2. PHOTOVOLTAIC ARRAY

The power of the array of each installation was determined by the technical coordination of the Project taking account of the needs of the user determined during the choice of the site and according to the standardized power steps of the generators. The output voltages of the arrays were as follows :

TYPE OF GENERATOR	OUTPUT VOLTAGE OF THE ARRAY
"400 Wp, D.C."	24 V
"800 Wp, D.C."	24 V
"800 Wp, A.C."	48 V
"1 200 Wp, A.C."	48 V

From the voltage and the power, the number of modules placed in series strings or paralleled can be derived automatically. The modules used for the programme are these of the French manufacturer Photowatt. Both monocrystalline PWP 400 and semi-crystalline PWP 402 silicon modules have been used for the generators. These two types of 40 Wp modules are fully accepted by the CEC's JRC Ispra as French state-of-the-art modules. They are also consistent with French regulation NF C 57-100. Their price for the programme is 8.7 ECU per peak watt (ex-works).

Thus the PV arrays are as follows :

TYPE OF GENERATOR	NOMINAL STORAGE CAPACITY
"400 Wp, D.C."	10
"800 Wp, D.C."	20
"800 Wp, A.C."	20
"1 200 Wp, A.C."	32

The typical selling price by the manufacturers for a 800 Wp array are :

- modules	:	6 660 ECU
- supporting structure	:	730 ECU
Total		7 390 ECU

Nevertheless, each installer uses his own type of supporting structure.

For instance, the 800 Wp photovoltaïc array of a generator in the Alpes de Haute Provence, Southeastern France (installation Nr.7) is divided in two subarrays of ten 40 Wp modules. Each subarray is made of two and a half series strings of four modules. From the junction box on the rear part of the panels leave three wires to the control command cabinet in the near by shelter.

The supporting structures are made from aluminium frame, bolted to the concrete foundations. The metallic parts are tied to the ground by copper wires.

For the programme, the structures of the array are designed in order to support the load of a 180 km/h wind.

The tilt angle of the arrays is generally 45° with regard to the horizontal plane.

For some generators, this angle is 60° facilitating the production of electricity in winter.

3. POWER CONDITIONING

The project offers two principal cases :
- D.C. generators,
- A.C. generators.

The power conditioning system, then, consists either of a simple battery-charging regulator, or an inverter in addition to it.

a) Battery size

The capacity of the battery has been calculated to allow the user three days storage on average, taking into account that it can only be run down to 70 % capacity, the speed of decharge and the energy efficiency of charging/decharging.

TYPE OF GENERATOR	NOMINAL STORAGE CAPACITY
"400 Wp, D.C."	8.4 kWh
"800 Wp, D.C."	14.4 kWh
"800 Wp, A.C."	16.8 kWh
"1 200 Wp, A.C."	28.8 kWh

b) Regulation of battery charging

The regulator allows charging to be maintained between 30 and 100 % of the battery's maximum charge. Each manufacturer uses his own means of regulation.

The state-of-charge of the battery is controlled by the battery voltage for all the generators. When over-charging occurs (2.4 V per cell), in some cases the array is disconnected, in other a floating current is maintained. If the voltage drops below 1.9 V per cell, usage is stopped in order that the solar panel can recharge the battery.

c) Inverters

Two 1 KVA inverters have been developped by two manufacturers SGTE and Faiveley, their performances are very similar :
- overload capability of 2 times the nominal power for a few minutes,
- sinusoïdal output waveform,
- electronic load detection (stand-by consumption 4 W, no-load consumption 55 W).

d) Back-up supplies

Some of the users owned a generating set. When their solar generator produces A.C. current, the thermal genset can be used as a backup through a reversing switch. In the opposite case, the photovoltaïc generator provides the permanent needs of the users and the genset is only reserved for the specific uses needing greater power.

4. DETAILS OF PV USER APPLIANCES

Every user consumes electricity differently. However, its most frequent uses are for :
- lighting,
- television and audiovisual appliances (including micro-computers),
- refrigeration,
- small domestic appliances (mixer, ...).

Other uses are :
- ventilation for the drying of medicinal herbs,
- low power machine-tools, etc ...

For all these applications, the user benefitted from the advice of the programme's Technical Coordinator and of the manufacturers in order that he supplies himself with equipment and economical electricity systems.

For instance, the following devices have been recommended :
- 13 W fluorescent lamp producing the same light as a 60 W incandescent lamp,
- television set with a 56 cm screen of 48 W consumption,
- 200 l refrigerator with standardized consumption of 0.5 kWh/day.

Unfortunately, for cost reasons, or because the user has an old apparatus, some appliances do not have such an economical consumption.

In fact, the more used appliances (2 or 3 hours a day) must be very economical, whereas it is less necessary for appliances only used intermittently.

5. MONITORING SCHEME

Two versions of instrument drawer have been developped, one for D.C. current generators, the other for A.C. generators.

The drawers are installed at the same time as the generators, and their use is explained to the user upon their acceptance. The instrumentation allows for easy data collection by the non-technical users who must send the data to the Technical Coordinator at the end of each month.

The drawers display the following measurements :

- in D.C. :
  - . "IG in kWh/m² : integration of the overall irradiation by a CIMEL pyranometer and its electronics,
  - . "Ahpv" in Ah : integration of the current generated by the photovoltaïc array,
  - . "Ahb" in Ah : integration of the current entering the battery,
  - . "Ahc" in Ah : integration of current used by the charge,
  - . "Ch" in h : display of the duration of generator usage,
- in A.C. version :
  - . "Aho" in Ah : integration of current entering the inverter and
  - . "kWhc" in kWh : integration of energy used by the user, by a watthourmeter.

Moreover, some clients' existing generator sets have been equipped with watthourmeters in order to know the energy they supply.

All these electromechanical counters must be recorded each week by the users who fill out a "counter-readings sheet".

All pyranometers have been calibrated by the Carpentras French National Meteorology Centre.

The "counter-reading sheet" is transcribed onto magnetic medium. A number of figures can be made, in particular :
- generator productivity,
- photovoltaïc array productivity,
- adequacy coefficient on site, taking into account the "quality" of generator sizing and the manner in which the generator is used.

In addition to the analysis made by the JRC, AFME has made two detailed studies :
- the first involves the search of the daily profile of electricity consumption by the reading of the consumption of the various appliances at 3 users' homes,
- the second involves the comparison of the quantitative data of consumption drawn from the "counter-reading sheets" with the opinions of the user regarding their own generator collected via interviews of eleven users.

The results of these studies will be available in the next weeks.

6. EXPERIENCE OBTAINED DURING INSTALLATION AND COMMISSIONING

Every installer or contractor chosen for the production of the programme's 40 generators has acquired his own experience during the installation and the starting-up of the generators. Several among them (in the French departments of Vaucluse or Alpes de Haute Provence for instance) have therefore had the opportunity to carry out their first photovoltaïc installation. These electricians have been helped in this task by the rules laid down for the programme. One of the latter's aims is the spread of photovoltaïc technology in the regions.

This experience has concerned mainly the safety rules which must be respected :

- grounding of the metallic structures,
- electrical insulation of the terminals of the cells and mechanical protection of the batteries,
- necessity of fuses to protect the various components of the generator,
- order for the connection of the systems, etc ...

In fact, the installation of the photovoltaïc generators presented no problems wherever the work was carried out carefully and according to accepted standards.

7. EXPERIENCE OBTAINED DURING OPERATION

Each user operates his own generator and must ensure that his consumption of electricity is in accordance with the generator production capacity and the weather conditions. So many cases arise between the following extremes : in the first, the user rapidly increases the number of his appliances, for instance by buying a second refrigerator and the generator appears underdesigned. In the other, the less well-off user has no money to buy equipment as rapidly as he wants, so his consumption is well below the capabilities of his generator and increases only slowly.

Then for two generators of the same size (400 Wp), one in Corsica, the other in the French Department of Drôme, the consumption in the winter period from the beginning of December 86 to the end of February 87 varies respectively from 170 to 260 Wh/day for the first user, to 480 to 680 Wh/day for the other without any connection with the actual composition of the families but related to their way of life.

Another main topic is the on-site efficiency of the photovoltaïc modules. It depends also on the consumption of the user. For the generator n° 28, the figure varies between 4.00 % and 7.90 % in the period beginning in September 85 and finishing in March 86. During this period, the energy consumption per square meter of modules and per day varies between 133.3 Wh and 262.5 Wh.

As for data monitoring, there are two main problems today. The first is that the users don't send their data sheets on a regular enough basis. The second is the breakdown of some amperhoursmeters thus preventing reliable data analysis.

INSTRUMENT DRAWER AND BATTERY STATE-OF-CHARGE REGULATOR

(Installation n° 31)

800 Wp, PV ARRAY

(Installation n° 24, Gard, France)

CONTRACT NR SE/Ø84/84

STAND - ALONE RESIDENTIAL PV-SYSTEM WITH MOTOR/GENERATOR BACKUP

H.K.KOETHE and E.ROESSLER
Fraunhofer Institut für
Solare Energiesysteme (ISE)
Oltmannsstr.22
D - 7800 Freiburg

Summary

The aim of the project is the economic operation of a stand-alone power supply system for a remote youth center in Orbey-Le-Schoultzbach, Alsace. The system consist of a PV-plant with a motor-generator back-up, driven by butane gas. The exhaust heat is utilized to cover part of the heat demand of the residence. Up to now, the PV-plant, motor generator, storage-battery, rectifier, inverter and the measuring-system have been installed and all components, except the monitoring system operate without problems.

Overvoltages of atmospheric or operational origin caused the malfunctioning and damage of the data acquisition system. While some data loggers have built-in overvoltage protection, our system had to be fitted with such devices. The reinstallation and protection of the data logger was finished at the end of March 1987 and the data acquisition period has been started again.

1. INTRODUCTION

The subject of the project is the demonstration of the economic operation of a stand-alone power supply system, consisting of a photovoltaic power plant (PV-plant) combined with a motor-generator, driven by butane gas. The effect of this combination should be

- saving maintenance and repair costs with the motor-generator by avoiding operation at low load and by covering low load requirements with the PV-plant.

- operation of the motor-generator at high efficiency and with a minimum of environmental pollution.

- using both electrical and thermal energy produced by the motor-generator.

During summer, the PV-plant should cover almost all load requirements. During winter, the motor-generator should be the main source of electrical and thermal power simultaneously.

As shown in the diagram Fig. 1, the electrical system consists of

- PV-plant with solar generator (PV-array) and power storage.

- motor-generator with back-up power via rectifier.

A control-logic unit will give priority to the PV-system at all times, if electrical energy is required by the consumer loads. The motor-generator will be started only if the battery is deeply discharged or if 3-phase current is required.

2. PHOTOVOLTAIC ARRAY

Originally the PV-array was to be mounted on the roof of the building, where the panels should replace the roof tiles. However, the permission for integrating the solar generator into the roof of the main building has not been given by the respective boards. Therefore, the solar generator has been mounted on a greenhouse-structure, erected at a distance of approx. 40 m from the main building. Nevertheless, the PV-modules cover only the part of the roof orientated to the south, and are mounted in the same watertight manner as previously planned for the building roof. The greenhouse-structure was constructed of steel, protected by a zinc coating.

The solar array field consists of 39 solar modules, type PQ 10/40/01, manufactured by AEG, with a total peak power of 1.48 kW. The main system feeding the battery and DC-loads consists of 38 of these modules.

One module provides the power supply for the data acquisition.

3. POWER CONDITIONING

If there is no need for 3-phase AC-current, the motor-generator is only used, if additional electrical power is required for recharging a flat battery. Thus, apart from this exception, the system will be controlled mainly by the state of charge of the storage battery. Consequently the control philosophy is as follows (also see Fig.1):

3.1 Normal operation:

The solar generator will provide the battery with DC-energy (EA) via the blocking diode. A shunt regulator will prevent the battery from being charged excessively by limiting the charging voltage to a level of 2.35 Volts per cell. The battery provides energy directly to the monophase inverter (EII) which covers the needs of the AC-loads (EIO).

3.2 Back-up operation:

* If the battery experiences a deep discharge, i.e. a discharge voltage close to undervoltage cut-off

or

* if 3-phase AC-current is switched on manually (ELAC2)

then the inverter will be turned off and the motor-generator will be connected to the consumer loads; hence the motor-generator will then

- recharge the battery via the rectifier (EG2, EG3)
- provide power to AC-loads (ELAC1)
- provide power to 3-phase AC-loads (ELAC2)

The plant will return to PV-power supply when the battery voltage reaches 2.2 Volts per cell on charging

or

if the 3-phase AC-requirements no longer exists.

4. DETAILS OF PV USER APPLIANCES

When planning the PV-system, the aim was to dimension the PV-plant and the storage-battery such that during the summer half year (15.03 - 15.09.), the average daily consumption could be covered. The dimensioning for operation in the summer allows a considerable reduction in the PV-plant and battery size compared to operation all the year round.

Hence a peak power of 1.44 kW at the maximum power point (MPP) resulted for the solar electric generator. For this configuration, two solar modules had to be connected in series, because a system voltage of 24 volts had been chosen for reasons of safety and practicability.

To guarantee long-term operation of the whole system without any significant maintenance, a lead acid battery of the type Varta bloc (Vb 2312), consisting of 24 cells (24 V) with a nominal capacity of 2 x 600 Ah, has been chosen. The advantage of this battery is its long lifetime of about 1300 cycles (processes of charging and discharging) with a discharging rate of 75% or about 4000 cycles with a discharging of 30%. Further more the Varta bloc battery has, compared to other acid lead batteries, a low self-discharging rate and a (relatively) high charging efficiency of about 95-98% (Ah - efficiency).

The battery can also be charged by the motor generator via a charging rectifier with a maximal power of 2.2 kW. This rectifier is fitted out with a semi-automatically controlled three-phase bridge-type rectifying circuit and charges the battery according to the current-voltage characteristic curve with a constant current of 80 A up to the voltage maximum of 28.2 V, then with reduced current and constant voltage.

To be able to use both, phase-controlled apparaters and electronically controlled motors, a sinusoidal inverter with a nominal power of 1.0 kW and a peak power of 1.4 kW has been installed.
As a result a relatively low efficiency of about 70% has to be accepted.
For this reason, the operating time of the inverter was minimized by operating as many consumer appliances as possible at 24 V.

Because of the PV system voltage of 24 V DC, there are three voltage levels in the Oberlin House, which are available from corresponding distribution circuits. For the lighting, only 24 V bulbs have been installed, which operate directly at the voltage level of the battery. Thus, all transforming losses caused by an inverter are avoided. Only specific rooms are equipped with plug sockets for consumers of alternating current, which are supplied by the 24 V DC/220 V AC - inverter.
To avoid idling losses, the inverter is only put into operation by a load detector when a 220 V - consumer is actually connected.
While the motor generator is running, the 220 V AC consumers are connected to the three-phase current-system in addition to the 380 V consumers. In this case, the inverter is switched off.

5. MONITORING SCHEME

All data necessary to define actual energy-flow and to calculate efficiencies of components, sub-systems and the total system, as well as turn-on times of the motor-generator will be recorded. Since some of the energies can be calculated from the others, not all of the currents etc. of the system will be measured. Thus, the following list of parameters is measured and calculated respectively (see Fig. 1):

5.1 meteorological data

- insolation (array plane) IA
- solar module temperature TC
- outdoor temperature TA

5.2 DC - energies

- solar generator current and voltage (EA)
- shunt current (EA 1)
- current to DC-loads (ELDC 2)
- current to inverter (EII)
- current from rectifier (EG 3)
- battery voltage

Calculations:

EC = EA - EA 1
ELDC 1 = ELDC 2 + EII
EB = EC + EG 3 - ELDC 2

5.3 AC - energies

- motor-generator output (EG)
- 3-phase loads (ELAC 2)
- rectifier input (EG 2)
- inverter output (EIO)

These energies are integrated by electricity meters equipped with pulse-output for recording.

Calculations:

EG 3 = EG 2 - ELAC 2
ELAC 1 = EG 1 + EIO - EG 2

All recording is performed by a data logger witch has 16 analog input channels and 7 pulse input channels with 12 bit resolution. Readings are taken every 2 seconds and average values calculated from these readings will be recorded every hour.

The data logger is able to take up to four plug-in memories with 8 kbyte or four memories with 32 kbyte. With one 8 kbyte-memory we can record data from 16 channels for about 20 days. The power supply for the memories is separated from the data logger and is provided by a lithium battery. An advantage of this data logger is, that it has a low power consumption of 0.5 W and an on-board lead acid battery. Therefore it was possible to use a separate solar module to power the data logger.

Reading the memories and calculating the energy flow is done by an IBM-PC using special equipment, which transfers the data from the memories via an RS 232 interface to the IBM-PC.

6. EXPERIENCE OBTAINED DURING INSTALLATION AND COMMISSIONING

The only problem that we had during installation was, that the respective boards of the french government did not give us the permission for integrating the solar generator into the roof of the main building. Thus we had to find another solution to install the solar generator. The result was that the solar generator now has been mounted on a greenhouse-structure at a distance of approx. 40 m from the main building. The disadvantage of this solution is the length of approx. 70 m of the circuits of the solar generator, the module temperature and the solarimeter to the control-logic, with the resulting electrical problems.

7. EXPERIENCE OBTAINED DURING OPERATION

During operation we had problems with sporadic peak voltages on the measured channels and some circuits of the control-logic. This lead to a partial destruction of the data logger and the shunt regulator of the system. The data logger is now fitted with a special overvoltage protection and we shall do the same with the shunt regulator. The measuring period restarted in March 1987 and there will be data available for ISPRA from this plant in May 1987.

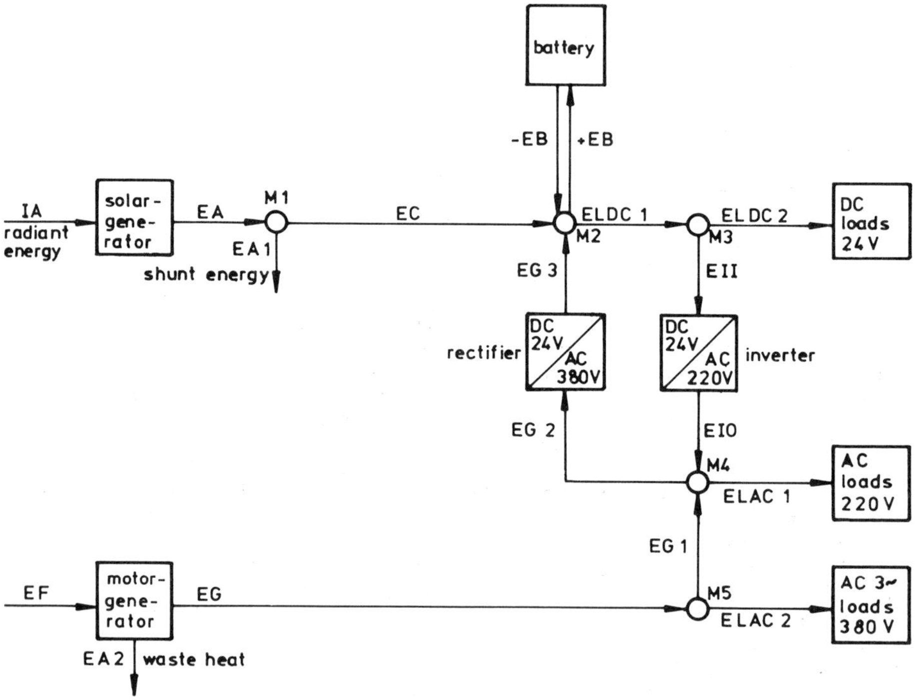

fig.1
energy flow of the power supply system

CONTRACT NO SE/507/84

SOLAR/WIND COGENERATION FOR URBAN RESIDENTIAL USE

A Dichler*, P R Wolfe*, R G Ardrey[o] and G Boyle[+]

* Solapak Limited, Factory Three Cock Lane. High Wycombe, Bucks HP13 7DE, UK

[o] Costain International Limited, Wokingham UK

[+] Open University Milton Keynes, UK

Summary

The **Solar/Wind Cogeneration Project** at Milton Keynes is now a highly visible demonstration of new generating technologies in operation. Its environmentally sympathetic design has made it an acceptable pioneer for alternative generation in an urban UK location.

Although highly innovative in terms of its application and realisation, the building blocks of which the system is comprised are based on proven technology. Safety and security of power supply have been paramount even where this conflicted with other priority features such as high efficiency.

The main equipment comprising solar arrays (both on houses and the energy management centre), wind turbine generator, power conditioning and monitoring equipment were installed on time and inaugurated in September. The remaining houses are scheduled for completion and occupation in the near future whereafter monitoring of system performance will continue until the end of the programme.

1. INTRODUCTION

It is the problem of storage which is the greatest obstacle to the widespread usage of 'alternative' forms of electricity generation. A mix of power sources, however, can go along way towards solving this problem, either by selecting various complementary local generator types or by coupling to a traditional grid system for storage.

The solar/wind project was conceived to establish how fuel-free generating systems can be made viable without having to provide excessive seasonal storage. Provisional analysis shows that the winter electricity demand in the UK is around double that of the summer demand. Wind availability is such that winter generation could be up to 5 times higher than in summer, while solar output would be up to 8 times higher in the summer than in winter. An appropriate mix of **complementary solar/wind** cogeneration devices should therefore be able to provide the required **summer to winter power profile.** Similar principles can be adopted with different ratios in other areas of the world.

This project involved the prototyping of such a solar/wind cogeneration system to supply electricity for residential application to nine houses in the United Kingdom. It will also be a test system to prove the concept of hybrid generation, intended for later commercialisation on a worldwide basis.

1.1 Current Status

The project was installed at an excellent demonstration site. Energy Park in Milton Keynes UK, which provides a showpiece for sympathetic and effective alternative generation schemes. The background to the project definition and its significance for future similar programmes was given in the original project proposal and other papers (10.1).

Installation of all major equipment was completed on time, despite the project not commencing in earnest until January 1986. the scheme formed part of the Energy World exhibition starting in August 1986 and was inaugurated by the Prime Minister Margaret Thatcher MP in September 1986 (see Fig 1). The house building programme is controlled by an outside developer to suit his commercial requirements. and this has extended the period to realisation of the full load, but general progress has been good.

Fig 1 Inauguration day.

1.2 Technical Scope of Project

The scheme involved the supply of the following major elements which are illustrated schematically in Fig 2 overleaf

A photovoltaic array totalling 4.6 KWp installed partly in the houses, and partly on The Energy Management Centre (which houses the battery. inverter control. switching and monitoring systems and also acts as a connection output centre for power to the houses).

A battery bank totalling 132 kWhr connected in two series strings of 110 volts.

An inverter rated at 20KVA based on a standard UPS design but modified for this project.

A wind turbine generator (WTG) rated at 22kW installed nearby.

Control and switching equipment regulating the power flow between the various elements and facilitating output to the houses and a backup link from the grid.

A monitoring system to record and analyse performance.

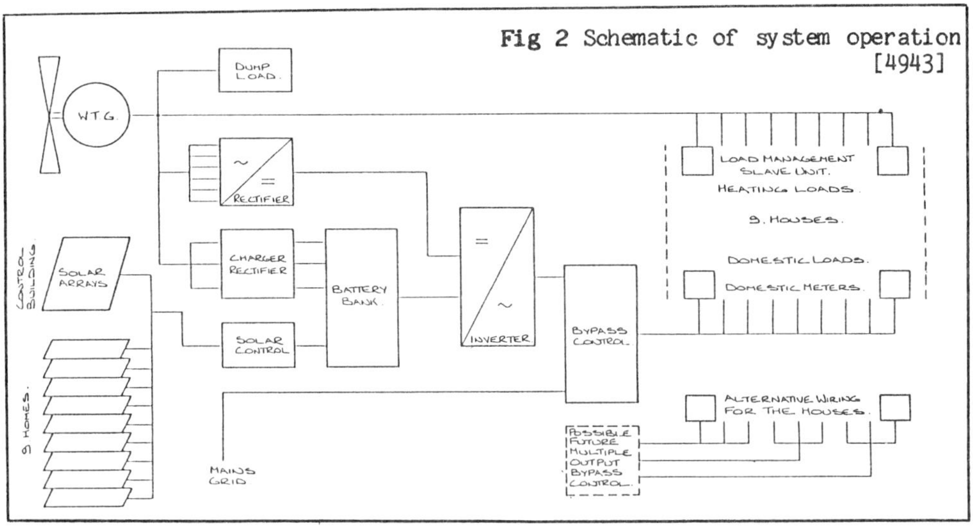

Fig 2 Schematic of system operation [4943]

The system operation has been described in a previous publication (10.2). Outline details of the main system elements is given in the following sections.

2. THE PHOTOVOLTAIC ARRAY

An important aspect of the project was the integration of the power plant and particularly the solar array into the house designs in a way which is acceptable to the surrounding environment.

Fig 3 Architects' impression of scheme

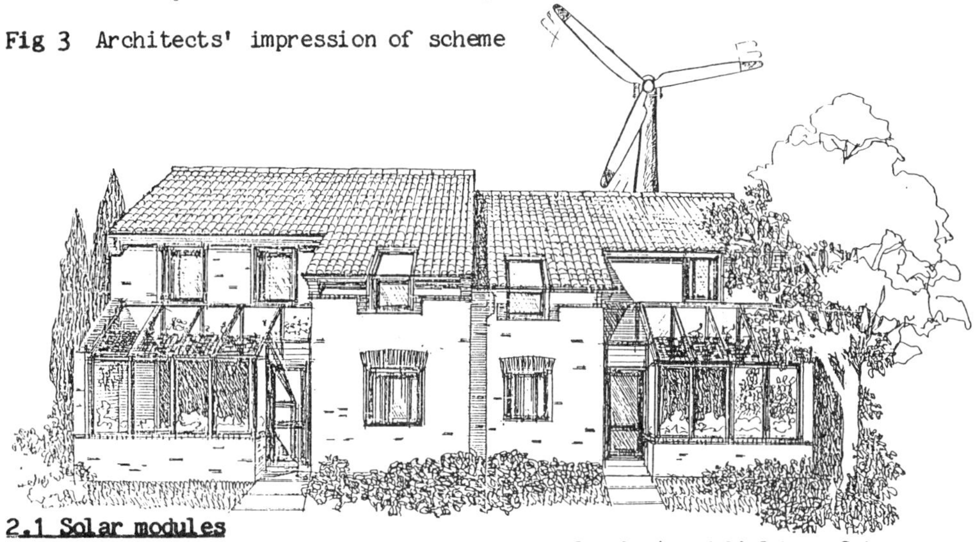

2.1 Solar modules

The solar array is connected as a total of 14 subfields. Subarrays 1 to 9 are fixed tilt and integrated into the conservatory structure of the 9 houses as shown on the architects impression (Fig 3). Subarrays 10 to 14 are mounted on the roof of the Solapak Energy Management Centre and facilitate seasonal adjustment of tilt by a novel latching mechanism Two solarimeters are also included on the roof structure.

Three different types of modules are used in the project, differentiated for the purpose of simplicity as follows:

	Wp	No of cells	Dimensions	
Type A	59	66	2540 x 642mm	} Cells fitted in 2/3
Type B	70	78	2540 x 661mm	} of the overall length
Type C	80	80	1100 x 875mm	

The Type C modules used on the Energy Management Centre are the supplier's standard module but mounted in double module blocks. Though type A and B modules are specially constructed to suit the roof dimensions of this project, they incorporate (except for the framing) all the same materials and manufacturing processes as type C modules. The cell types, method of interconnection, superstrate and encapsulation are all identical to the standard. The only mechanical difference is in the mounting arrangements where the modules rely on the structure of the houses themselves for their mechanical support. Module performance prediction is based on the certified data provided by the supplier. These modules are mounted in all respects like standard glazing panels and use identical design principles which are now well proven over many decades.

Because of the large glass sizes used in the type A and B modules to achieve the desired architectural integration, lamination was not feasible so a construction where cells are encapsulated in a silicon resin was selected. This has been proven over many years in the field. For future projects of this type it would be preferable to develop a construction which lends itself more readily to high volume, low cost production.

The total cost of the solar modules and interconnections was about £32,000 representing £7.15/Wp. There were further additional inputs relating to the development of these special units totalling some £8,000. The variable structure for the Energy Management Centre cost around £2,000, while the support structure provided by the houses was effectively free.

2.2 Solar Controls

The general control philosophy and central power conditioning equipment control are detailed in Sections 4 and 6. The solar DC controls are breifly described below.

For safety reasons and to make the solar array mechanically and electrically manageable, a DC operating voltage of 110 volts was chosen, so that two sets of DC controls (and batteries) are fitted, one 'stacked' above the other to derive the 220 volts DC required by the UPS system and load.

The controls are configured so that each control 'half' can accept up to 6 strings each at 110V DC, fed separately to 6 regulator switches. Each regulator switch employs a series connected FET device for control, and a silicon blocking diode to prevent reverse current flow. A novel feature of the Regulator Switch is that it employs an opto-isolator in the control drive path to the FET gate, so the control electronics are completely isolated from the high battery voltage.

Each regulator switch is driven separately from a regulator control card via 6 control drive lines. It is the regulator control card that monitors the battery (via a separate pair of voltage sense leads to avoid cable volt drop errors) and after temperature compensation instructs the regulator switches to operate. An interesting feature of the control card is its ability to cause the regulator switches to operate one after the other rather than all together. This cascade mode, enhances the battery charge acceptance and further reduces any risk of r.f.i. due to high rates of change of current in cable conductors.

As a back-up to the Regulate function, a Battery Monitor Card constantly compares the battery voltage against a known, and very stable internal reference voltage. If for any reason the battery exceeds preset limits, high or low voltage alarms are generated, and if these are maintained for any length of time corrective action is taken. In the case of a High Voltage Alarm, action would be by manual intervention and investigation into the fault. A Low Voltage Alarm, however, will cause the housing load to be transferred from the Solar/Wind Scheme to the local grid supply by operating the static switch (See Sections 6.3 and 6.4).

3. STORAGE AND BACKUP POWER SUPPLIES

Storage is incorporated to provide a total of one day's average expected load on the system.

The battery chosen is a tubular low loss flooded cell. The total battery comprises 110 cells each of nominal capacity 600 Ampere hours at the 10 hour rate. The float voltage recommended is 2.23 V DC and was adopted initially. However, given the cyclic duty expected and the variability of the input energy, upward adjustment of the float voltage is anticipated. The extent will only be known as the system becomes more fully operational and of course the supplier will be consulted.

The battery is split into two electrically identical halves and receives charge from the solar arrays into each of these halves in equal proportion. Charging from the Wind Turbine, however, is accomplished at the 220 Volts level.

There is an emergency backup connection to the local grid which can be activated in addition to the bypass switch in the power conditioning unit. The system design and the storage element detailed above should minimise the need for contribution from the local grid, but this will only be known when the full load is in normal operation and the load/generation match is assessed.

4. POWER CONDITIONING UNIT

The inverter is based on a standard UPS system that has had some modification to suit this application. It should be noted that the modifications relate to operation, external interface, and safety considerations rather than to internal improvements in operating efficiency. The output of the UPS system is fed via a no break static switch to the housing load. When the load exceeds the capability of the solar/wind system it is transparently switched to the local grid supply. Both the part which connects power from the inverter, and that which connects power from the local grid (via the 65KVA isolating transformer) are monitored. In the event of both passing current (under fault conditions) the mains grid is disconnected and the inverter shut down. Various manual controls have been built into the UPS for maintenance purposes.

The power conditioning unit further provides a three stage regulated battery charger rated at 30 amperes maximum which is the first load presented to the WTG. See Section 6.6.

5. WIND TURBINE GENERATOR AND LOAD EQUIPMENT

The design and provision of the wind turbine generator and the overall integration of the total system with the load equipment has been a major part of the project. The former will be described only briefly, however, as this paper is intended to address mainly the solar systems aspects.

5.1 Wind Turbine Generator (WTG)

The selection of the wind generator was heavily influenced by the required system reliability and safety factors since this is the first urban installation in mainland Britain. The suppliers' experience with grid interaction was also a consideration.

The originally specified capacity was 20kW nominal. In fact a unit with a rated capacity of 22KW was installed and designed especially for a low wind speed area. This was achieved by modifying a standard unit, increasing the swept area with a specially extended wing mounting and by careful choice of wing profile. The actual output of course varies from the rated figure depending on wind speed and in early tests in moderately strong winds a peak output between 30 and 33 kW has already been experienced.

At the Milton Keynes site wind records show that there is often sufficient wind for the WTG to continue to run once started, but that starting wind speeds are less frequent in occurrence. A starter motor was therefore added to the WTG that will accelerate the wings to normal running speed within 10 seconds of being triggered by the anemometer. It is anticipated that this addition will increase the annual output of the WTG by approximately 30% over the level otherwise expected.

5.2 Load equipment

The system has been designed primarily to meet standard residential power requirements. For the users' point of view it will perform exactly like a mains grid supply.

The only special facilities in the houses are immersion heaters which are be activated by the WTG control when surplus power is available from the WTG. These heaters are supplied direct from the WTG in preference to being supplied from the inverter, since precise waveform and frequency is not critical for these devices.

6. CONTROL AND SWITCHING EQUIPMENT

The control regime is illustrated in Fig 4 overleaf and summarised as follows:

6.1) The batteries are permanently charged from the solar array.

6.2) The output to the houses is normally fed through the inverter taking an input preferentially from the WTG through the rectifier when windspeeds allow or otherwise from the battery as required.

6.3) In case of insufficient power from the WTG to support the load, accompanied by inadequate battery state of charge, the grid standby is activated, and the load is transferred without noticeable break.

6.4) In the event of grid failure when standby mode is selected the battery S.O.C. override is reset to a lower level so that the inverter continues to support the house loads.

6.5) In case of overload on the inverter the grid is also activated.

6.6) The WTG control maintains rotational speed by loading and unloading the turbine electrically. The loads are applied in the following sequence:

- i) Regulated battery charger (3 stages)
- ii) Inverter (with total of 6 stages of incremental load) deriving output to the housing load.
- iii) Immersion heater loads in 9 houses (first stage dump).
- iv) Standby multistage dump load (if required to prevent overspeed)

This algorythm has been refined to provide, using proven equipment, an optimum match of systems efficiency, safety and security of power availability.

Fig 4 System Control Algorythm (SP0126)

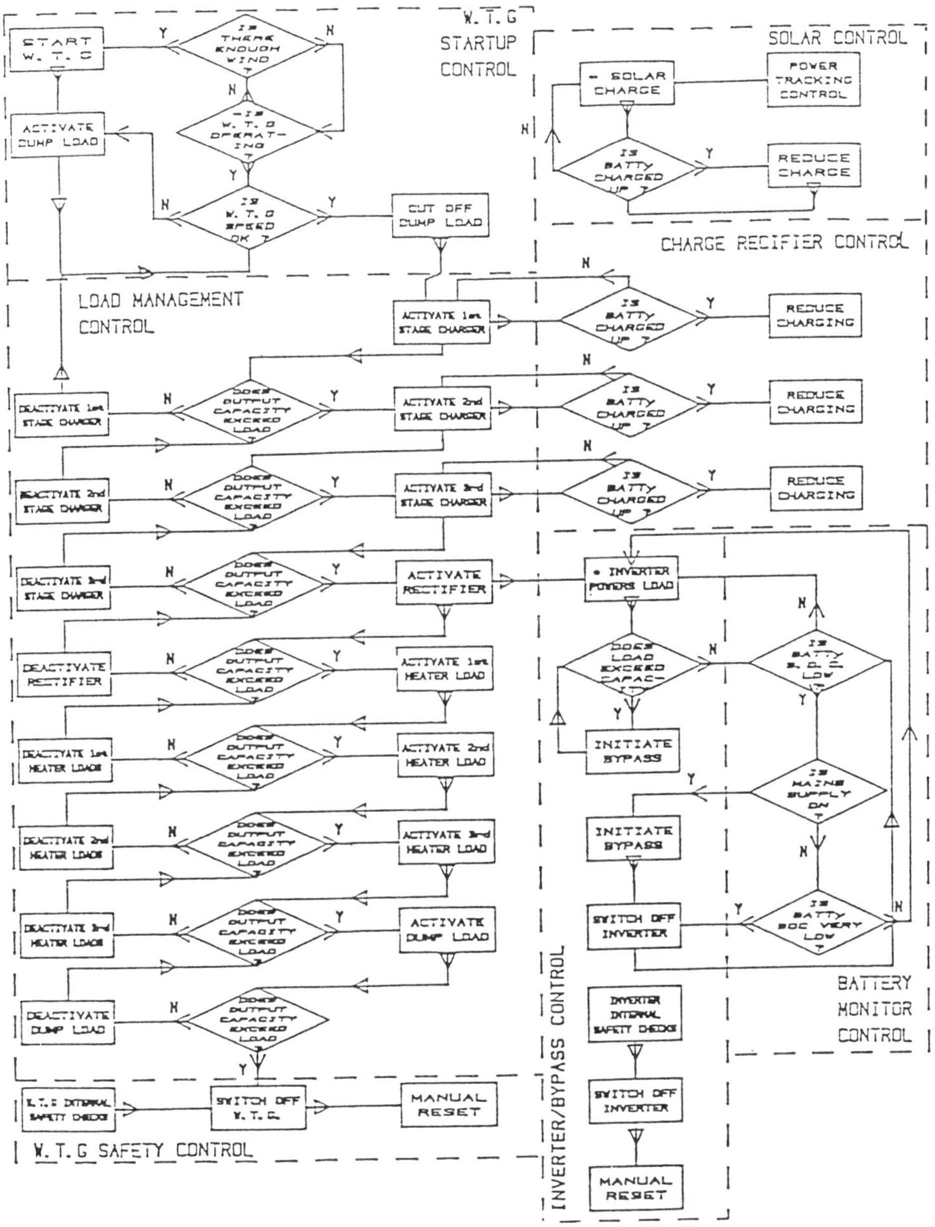

7. MONITORING

In order to carry out performance monitoring of the various systems and sub-systems involved in the project, in accordance with the guidelines of the Commission of the European Communities' (CEC) Ispra Research Centre. various sensors, instruments, data acquisition and computing equipment have been installed at the project site. in the configuration shown in Fig 5.

Fig 5 Block diagram of monitoring scheme [SP0178]

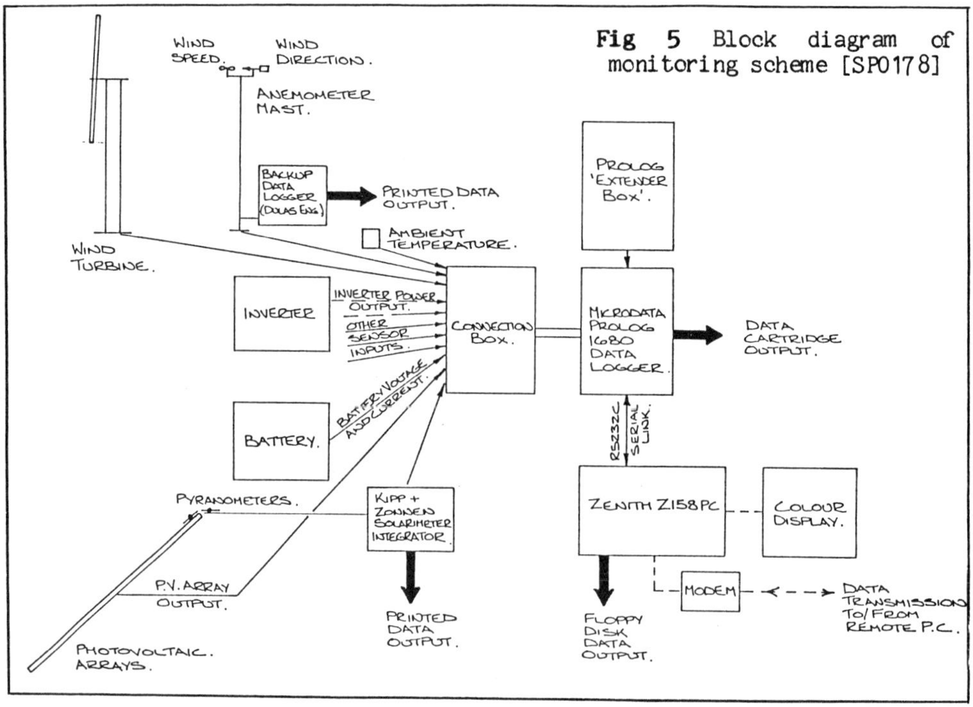

7.1 Monitoring parameters, frequency and data logging policy

Sensors which measure the main meteorological parameters (solar radiation, wind speed and direction, ambient temperature etc.) and the main energy flows within the system (energy output of wind turbine. photovoltaic array output. energy delivered to loads, etc) have been installed.

Data inputs to the prolog data logger are scanned at regular intervals (currently every 10 minutes) and the results of each scan are recorded on a magnetic tape cartridge in the ECMA-46 standard data format. The parameters measured are detailed in Fig 6 overleaf which also indicates sensor types.

7.2 Data analysis

In order to facilitate data recording, backup and analysis, the information acquired by the Prolog system on each scan is also transferred, via an RS232C interface. to an IBM PC compatible microcomputer (a Zenith Data Systems Z158) and placed initially in a 'raw data' file on the computer's 20Mbyte hard disk. Every 24 hours, the 'raw' data is processed to create additional files, one for each day. in a 'spreadsheet' format compatible with the Lotus 'Symphony' software package; this enables the data to be analysed much more easily than in its 'raw' form. (The 'raw' data is. however. always available if required.)

Fig 6: Channel Allocations. Parameters Measured and Sensors Used

Prolog Channel Number	Parameter	CEC Symbol	Sensor	Comments
00	Insolation (global. horizontal)	IG	Kipp & Zonnen pyranometer type CM11	Prolog integrates solar radiation (kWh) during scan interval.
01	Insolation (array plane)	IA	Kipp & Zonnon pyranometer type CM5	Prolog integrates solar radiation (kWh) during scan interval.
02	Wind Speed (10 height)	VM	Vector Instruments digital anemometer	Prolog counts total 'run of wind' (m) during scan interval
03	Wind Speed (overflow from channel 02)			
04	Wind Direction	VD	Vector Instruments analog wind vane	Prolog reads instantaneous wind direction on each scan
05	Outside shaded temperature	TA	Platinum resistance thermometer	Prolog reads instantaneous ambient temperature on each scan
06	Photovoltaic array temperature	TC	Platinum resistance thermometer	Prolog reads instantan- array temperature on each scan
07	Photovoltaic array energy output	EA	Custom-built integrating energy transducer	Energy transducer provides integrated energy (kWh) reading
08	Photovoltaic array power output (overflow from channel 07)			
09	Array current (positive side)		Current transducer	Prolog reads instantaneous current on each scan
10	Array current (negative side)		Current transducer	Prolog reads instantaneous current on each scan
11	Battery voltage (positive side)		Direct measurement by Prolog	Prolog reads instantaneous voltage on each scan
12	Battery voltage (negative side)		Direct measurement by Prolog	Prolog reads instantaneous voltage on each scan
13	Array voltage (positive)		Voltage tapping (direct measurement) from energy transducer	Prolog reads instantaneous voltage on each scan
14	Array power output (positive side)		Power output tapping on energy transducer.	Prolog reads instantaneous power output on each scan
15	Wind Energy (Net output to system)		Opto-switch scanning disk on kWh meter	Prolog records total energy (kWh) during sacn interval
16	Wind Energy (to dump load)		Opto-switch scanning disk on kWh meter	Prolog records total energy (kWh) during scan interval
17	Wind Energy (Gross turbine output)	EG	Opto-switch scanning disk on kWh meter	Prolog records total energy (kWh) during scan interval

Software has been written to process the 'raw' data into the format recommended by the CEC. This data can be copied from the hard disk on to an IBM format floppy disk conforming to CEC requirements.

The use of the Lotus 'Symphony' package enables colour graphic displays of data trends and relationships to be created with ease on any IBM compatible personal computer.

The whole monitoring procedure is to be described more fully in a separate paper (10.4).

8. PROJECT MANAGEMENT. INSTALLATION AND COMMISSIONING

For the purpose of these procedings, this document has concentrated largely on the challenge for technical realisation of the project. The organisational and management aspects, however. proved to be of equal. if not of greater importance. Completing the installation to a demanding timetable and to the satisfaction of a large number of participants proved to be a major undertaking.

8.1 Project management

The main participants have been identified separately (10.5). Essential to the project's success, however, were inputs in project coordination from Costain International Ltd, in planning approvals from local authorities, Government Departments and MKDC, and in monitoring from the Open University.

The main systems design and integration was supervised by a team comprising Solapak (DC equipment supplier), IRD (the WTG supplier) and Powernetics (power conditioning supplier) as well as the project coordinators.

8.2 Installation and Commissioning

Installation and preliminary commissioning proceded very well to timetable despite the short lead times. The main problems turned out to be administrative in such areas as planning consents and general co-ordination with the overall Energy World Exhibition. However with goodwill all round these problems were eventually overcome.

The installation in particular was a substantial engineering job in its own right.

9. SYSTEM OPERATION

It is still early in the monitoring phase and detailed operational data will be available later in the programme. Several operational factors are already significant, however.

9.1 Enviromental monitoring

Because this is the first project of its kind in an urban enviroment in the UK (and possibly Europe), much attention is being paid to its enviromental acceptability. This applies not only to system design referred to above. but also operation.

Accordingly Solapak are participating in an enviromental monitoring study relating mainly in respect of the WTG which will assess two main factors:

Noise - with specific reference to the inhabitants of nearby house.

Public acceptance - both from the point of view of passing motorists and local residents.

9.2 Demonstration and education

The important role of this project for demonstrating the potential ofalternative generation and educating the public has been given high priority.

It formed an important part of the main Energy World Exhibition and has been visited by many overseas delegations (including IEA), government ministers and members of the public.

The equipment has been laid out and labelled to be as informative as possible and graphic outputs from the monitoring are designed to meet similar aims.

9.3 Ongoing system management

One of Solapak's objectives for the project was to ensure that provision was made for householders deriving electric power from the scheme to have participation in its operation. It was therefore decided to create a separate management body, with all householders as members together with a representative from Solapak and from the developers (Persimmon Homes). This body would ensure the householders had adequate representation in system operation and the ability to provide maintenance by entering into agreements with the main equipment contractors for the service life of the scheme. The management body will also be responsible for the correctness of operation of the scheme and provide management and financial services as necessary.

9.4 Provisional results

The houses which form the load for the system are being occupied progressively and are not expected to be fully inhabited until later in the year. Until then accurate profiles of load consumption (and therefore systems analysis) is not possible.

The monitoring system has however been in operation for some months and is already providing valuable climatic and performance data which enables operational simulation to be carried out relating to expected power generation. Further details are due to be published in due course (10.4).

10. REFERENCES

(10.1) 'An Integrated renewable energy system for a small conservation conscious community'. G A Boyle. Proc. Solar World Forum. 1981.

(10.2) 'The Solar/Wind project'. P R Wolfe. Proc. Energy World Business Week Conference. 1986.

(10.3) 'Solar/wind power project' Technical Report No. 1; Project SE/507/UK. A Dichler. 1986.

(10.4) 'Design and operation of a wind-photovoltaic cogeneration system for a group of houses'. A Dubois, G A Boyle and A Dichler. (proposed for) Proc. ISES Solar World Congress. 1987.

(10.5) 'Electricity from the Sun and Wind'. Solapak Publication on the Solar/wind Project. 1986.

Contract No. SE/520/84

INTEGRATION OF A PHOTOVOLTAIC SOLAR GENERATOR IN AN EXISTING RESISDENTIAL SYSTEM

G. T. WRIXON and S. MC CARTHY
National Microelectronics Research Centre,
University College, Lee Maltings,
Prospect Row, Cork,
Ireland.

Summary

The Rosemount Domestic Photovoltaic System is a 5.5kWp PV system which is used to supply electrical energy to a domestic residence at Rosemount, Farnanes in County Cork, Ireland. The system is connected directly to the utility grid and no batteries are used. A maximum power point tracker is used to ensure that all available PV array energy is converted into useful AC energy. The complete system operates automatically and a specially developed monitoring system is used to record data for analysis at the Joint Research Centre at Ispra, Italy. The system began operating in December 1986 and to date no major faults have occurred. The system is easy to use and it provides the user with real time data of the system performance on a computer terminal.

1. INTRODUCTION

The Rosemount Domestic Photovoltaic System consists of a 5.5KWp PV system which supplies electrical energy to an existing residential system. Figure 1 shows the overall layout of the system. It consists of a 5.5kWp PV array, and a 7kVA line-commutated inverter. No batteries are included. The output of the inverter is connected directly to the utility grid. The PV array is located on the roof of an adjoining shed (Figure 2) and the power conditioning equipment is located under the shed. The meteorological equipment and lightening protection equipment are also located on the roof of the shed. Power lines and signal cables are fed from the shed to a room at the rear of the domestic residence. In this room the power lines are connected to the utility grid and the signal cables are connected to the microcomputer system. The system manual, log book and spare parts are also stored in this room.

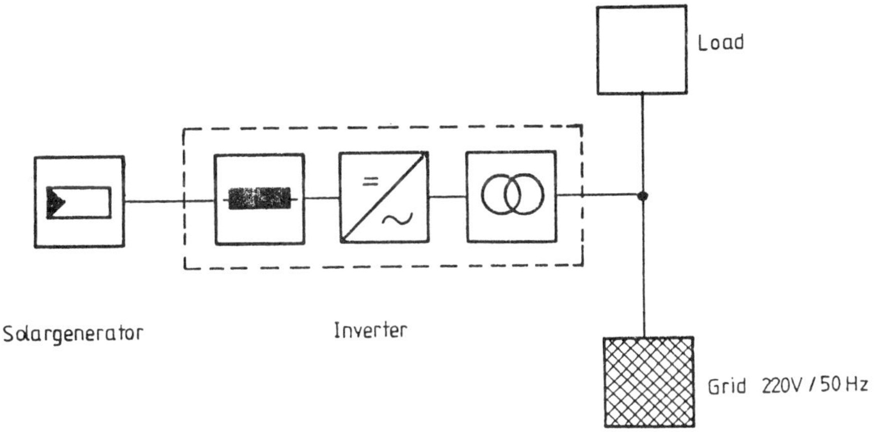

Figure 1. Block Diagram

Figure 2. PV Array

2. PHOTOVOLTAIC ARRAY

The PV array consists of 144 modules (type PQ 10/40/01) mounted on the roof of a shed. (Figure 2) There are 9 parallel strings each with 16 modules connected in series. The modules are mounted on timber supports and are held in position using special fastners (Figure 3). The timber is a softwood which was treated with a chemical preservative. The main technical data for the array is listed in Table 1.

Each module consists of forty 10 x10 cm^2 multicrystalline silicon solar cells and are fully qualified for the EEC specification EUR 501. The square cells are approx. 400 um thick and are doped n on p using a diffusion process. The front and rear contacts are made from aluminium and the anti-reflective coating is made from Ti Ox. Two shunt diodes, each bridging 20 solar cells, are integrated into the frame of each moudule.

The total cost of the PV array was 37714 ECU i.e. 6.82 ECU/Wp

Figure 3. The modules are mounted on timber supports and are held in position by special fastners.

Table 1: Technical Data of PV Array

Rated Power	5.53 kWp
Module Type	PQ 10/40/01 (AEG)
Number of Modules	144
Umpp	280V
Impp	19.8A
Isc	21.7A
Voc	358V
	(AM 1.5, $1000 wm^{-2}$, 25°C)

3. POWER CONDITIONING

Power condition is achieved using a 7kVA line commutated inverter and a MPPT controller. Figure 4 shows the layout of the power conditioning system and Table 2 is a summary of the technical data of the inverter. The electrical energy supplied by the solar generator is fed to the inverter via a fuse (Fl) and a choke (Ll). The capacitor Cl is used to smooth the DC voltage in order to operate the solar generator at MPP. AC energy from the inverter is fed to the utility grid and loaded through the transformer Tl. A capacitor (C2) is placed between the inverter and the transformer to reduce the reactive power supplied by the utility grid.

The inverter is line-commutated, therefore, when the utility grid fails the inverter automatically shuts down. Also if the utility grid voltage decreases the inverter automatically shuts down. The power conditioning system was designed to meet the safety requirements of the Irish utility grid (Electricity Supply Board).

During the night time the inverter is switched off using a time switch (Kl and K2 of Figure 4 are open). During the day time Kl and K2 are closed and the inverter operatates under MPP. A manual switch can also be used to turn off the inverter.

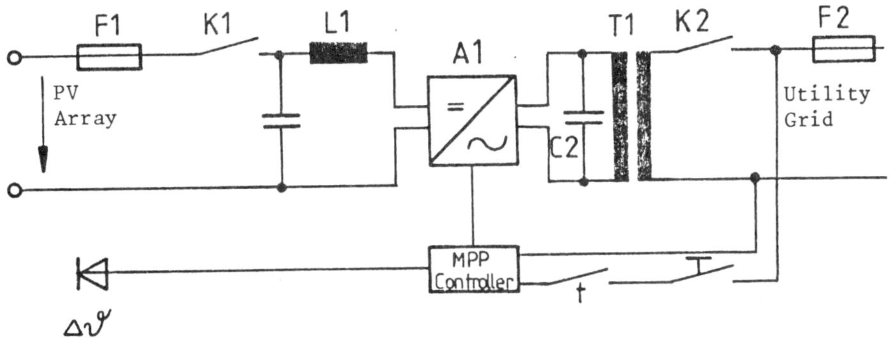

Figure 4. Inverter System

Table 2: Technical Data of Inverter

Type	MINISEMI (AEG)
Rating	7kVA
I max	24A DC
Maximum input voltage	280V DC
Output voltage	220V single phase 50Hz
Dimensions	600 x 1800 x 500 (WxHxD)
Weight	250 Kg

3.1 MAXIMUM POWER POINT TRACKING (MPPT)

MPP is implemented using a special control board located in the inverter cubicle. This board allows automatic MPP or manual MPP (for testing). Temperature compensation of the MPP is implemented using the temperature of a solar module. The temperature is measured using a silicon diode which is in thermal contact with the rear of a solar module.

The MPP card also performs automatic start-stop. Automatic start up begins when the PV array voltage exceeds a predefined threshold (Vhigh) and automatic shut-down begins when the voltage is less than a predefined threshold (Vlow).

4. DETAILS OF PV USERS APPLIANCES

The main appliances are washing machine, dish washer, fridge, freezer and lighting. During the summer months the mean daily load demand is 12kwh and during the winter months it increases to 15kwh. (Domestic heating is done using oil burner). Before the installation of the PV system the equipment was used at irregular intervals. Since the PV system was installed some activities such as washing is done around midday. Special eutectic plates have also been ordered so that the freezer can be operated during the day and turned off during the night.

5. DATA MONITORING SYSTEM

The data monitoring system uses a standard CMOS-microcomputer system (CSS-SYS 800) and a standard programming language CP/M 80. The system layout is shown in Figure 5. Data is inputted through 16 analog channels, each with a 12 resolution. The calculated values of energy are stored on standard 5¼ inch floppy disk. The complete system is mounted on a 19 inch rack and data can be monitored on a computer terminal using an RS232 interface.

The recording format of the data is designed to the "Guidelines for Photovoltaic System Monitoring".(1)

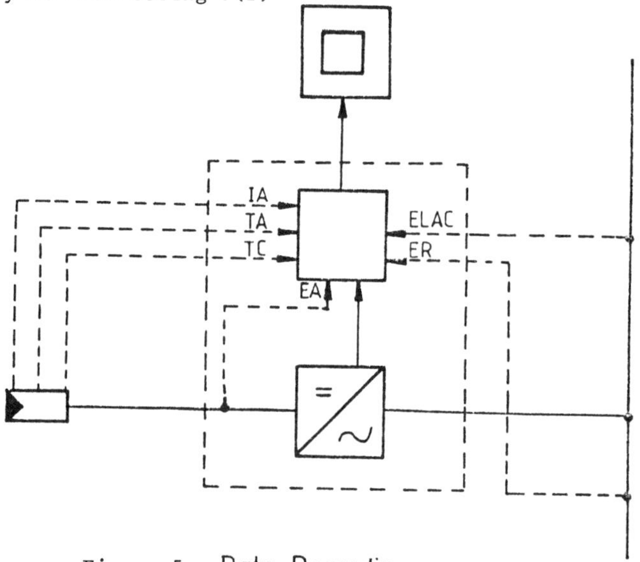

Figure 5. Data Recording

5.1 MONITORED DATA

Two types of data is monitored

- meterological data
- system data.

The meteorological data consists of ambient temperature which is measured using a thermocouple and planar irradiation which is measured using a reference solar cell.

The system data consists of:

- PV array current and voltage. These are used to calculate the PV array energy.
- AC energy to/from the utility grid.
- AC energy to the load.

Each measuring point is read every 10 seconds and the mean value is calculated every 10 minutes. Every 60 minutes the mean value for the hour is calculated and is stored on the floppy disk.

The measured data and the recorded data can be monitored on the computer terminal. This facility is used to examine the recorded data and to calibrate the measuring points. Measuring points can be calibrated by comparing the transducer signals (using a digital volt meter) to the values monitored on the computer terminal.

Each measuring point has two parameters defining the measured value i.e. scaling factor and zero offset.

$$\text{Measured value} = \text{scaling factor} \times \text{signal} + \text{offset}$$

These parameters are stored in a file and can be changed if calibration errors are detected.

Data recording began in December 1986 and to date four months data is available. To date no software has been developed to analyse the data.

6. EXPERIENCES GAINED DURING INSTALLATION AND COMMISSIONING

The main experience gained from this project was the designing of the system to meet the safety requirements of the utility grid. The main criterion was to ensure that the system was automatically disconnected from the utility grid when predefined limits of voltage and frequency were exceeded. Equipment to implement these safety requirements cost 2800 ECU but in this project the safety requirements were incorporated into the systems design. In future projects these design features can be incorporated as standard design features resulting in a saving to the user.

7. EXPERIENCE DURING OPERATION

The PV system began operating in December 1986 and after 4 months operation no major faults have been detected. From the users point of view the system is easy to operate and maintain. A set of operation and maintenance procedures have been developed for the system and these are kept at the NMRC.

These procedures consist of monthly manual inspections of the system and half yearly detailed inspection. The detailed inspection will measure the I-V curve of the PV array and will test the calibration of the measuring equipment.

8. CONCLUSION

The Rosemount domestic photovoltaic system is a good example of how experiences from a PV pilot plant are incorporated into a PV demonstration project. In this case the experiences from the Fotavoltaic Project are used in the system design and in the operation and maintenance of the system.

The system is designed using standard components i.e. mini semi inverter, Intel micro computer system, CP/M programming language etc. This will help overcome problems such as availability of spare parts and shortage of personnel with experience in hardware and software design.

The system design is very well documented. This will avoid problems which occur with changes in project personnel.

This project will provide valuable information to examine factors such as:

- the reliability and performance of grid connected PV systems
- the feasibility of using PV systems to supply the energy demands of domestic residences at nothern European latitudes.

REFERENCES

(1) BLAESSER, G. and KREBS, K. Guidelines for Photovoltaic System Monitoring. Joint Research Centre, Ispra Establishment, Italy.

(2) WRIXON, G. T. and MC CARTHY, S. Field Experience of a 50kWp Photovoltaic System on Fota Island. Proceedings of the 6th E.C. Photovoltaic Solar Energy Confernece held in London, April 1985.

(3) MC CARTHY, S., WRIXON, G. T. and KOVACK, A. Two Years Operating Experience of a 50kWp Photovoltaic System. Proceedings of the CIBSE/ASHRAE Conference held in Dublin, Ireland, September 1986 (129:135).

CONTRACT NR. SE/573/84
INTEGRATED ENERGY SYSTEM FOR POWER SUPPLY OF AN ISOLATED FARM LOCATED ON LIGURIAN APENNINES

C. MACCIO' and G. PASTORINO
CESEN - Via Pieragostini 50 - Genova - Italy

Summary

The aim of the project was the realization of an integrated energy system for power supply of an isolated farm located on Ligurian Apennines, not connected with the national grid, utilizing a photovoltaic solar system (2100 Wp) together with a micro-hydro equipment (350 Wel).
The P.V. system was installed in September 1985 and has been running since then with no remarkable problem.
On the other hand, as regards the micro-hydro system, the implementation of the project was not allowed by local authorities. As a consequence a pre-existing diesel generator is sometimes operated to meet the farm power requirements.
Detailed operating data will be available as soon as the data acquisition system will be put into service.

1. INTRODUCTION

The aim of the project was to meet the power requirements of a agricultural - touring centre named "Prato Rondanino" and located on a remote part of the Ligurian Apennines.

The centre is not supplied by the ENEL's grid and consists of three country houses: one of them is used for tourist lodging and is closed in winter season. The other ones are inhabited by resident people.

An integrated energy system was foreseen, combining two different energy technologies relevant to new and renewable energy sources: solar photovoltaic and hydroelectric.

The photovoltaic part of the energy system was installed in september 1985. It consists of the following main components:

- photovoltaic array 2.1 kWp
- electrochemical battery 42 kWh
- inverter 3.5 kVA
- interface and control board

The above P.V. system has been running since the erection without any remarkable problems, feeding the loads of the centre with the backup of a pre-existing 10 kVA diesel generator.

On the contrary the hydroelectric part met the veto of the local authorities owing to municipal cognizance problems as well as the opposition of the neighbouring water users.

So the present operating condition (PV system + diesel backup) has to be considered as the definitive one.

The installation of the data acquisition system is now under completion: starting from next month (May 1987) the plant will be continuously monitored for a two years period.

2. PHOTOVOLTAIC ARRAY

The photovoltaic field is composed of n° 60 P.V. modules type ANSALDO AP 35 HD, connected in a proper series-parallel configuration, with relevant protection diodes and mounted on a galvanized steel supporting structure, providing an adjustable tilt angle between 10° and 60° on the horizontal surface according to the season.

The main figures for the P.V. field are in the following table:

- No. of P.V. modules AP 35 HD	60
- Total peak power	2100 Wp
- No. of series strings	6
- No. of P.V. modules in series	10
- No. of supporting structures	3
- No. of P.V. cells	2160
- Terminal rated voltage	120 V
- Rated open circuit voltage	200 V
- Rated short circuit current	15 A

The average daily values of global solar radiation on the P.V. modules surface and the appraised output of the P.V. field are listed in the following table:

Month	Tilt Angle	Solar Radiation	Appraised P.V. field output
January	60°	2.06 kWh/m^2 day	28.4 Ah/day
February	40°	2.37 " "	32.7 "
March	40°	3.18 " "	43.9 "
April	10°	3.80 " "	51.3 "
May	10°	4.48 " "	60.5 "
June	10°	4.91 " "	65.6 "
July	10°	5.08 " "	67.8 "
August	10°	4.47 " "	59.7 "
Semptember	40°	3.70 " "	49.4 "
October	40°	2.84 " "	38.3 "
November	60°	1.95 " "	26.9 "
December	60°	1.82 " "	25.1 "
		"	
Yearly average	-	3.39 " "	45.8 "

3. POWER CONDITIONING

An electric control board provides the interfaces of the storage system with the inverter, the P.V. field and the pre-existent diesel generator which can also supply the loads directly.

The control system is allocated in the control board and has two main purposes:

- the regulation of the battery charge from the P.V. field
- the control of the enable-disable commands for the inverter according to the charge level of the battery.

Lead-acid batteries for stationary installations has been adopted, providing both a high charge/discharge efficiency and a very low self-discharge rate.

Even the maintenance demand (distilled water fill-up) is scarce; as above mentioned, batteries are protected from overcharging and excessive discharging by the control system.

The main features of the storage system are listed below:

- Rated voltage	120 V
- Rated capacity (10 h discharge)	350 Ah
- Energy capacity	42 kWh
- Number of elements	60
- Configuration	Series connection
- Electrolyte specific weight	1.25
- Size of one element	210x220x415 mm
- Weight of one element (wet)	30 kg

As regards the A.C. power supply to the electric loads a 3.5 kVA inverter was installed; its main features are summarized in the following table:

- Type:	Pulse width modulation transistor
- Output:	Single phase - 220 V - 50 Hz
- Input:	120 Vdc (100 ÷ 150 Vdc)
- Efficiency:	Full load - 90%
	Half load - 85%
- Protection:	Input overvoltage, output short-circuit
	Output overload
- Rated power capacity:	3,5 kVA (at cos φ = 0,8)

A second inverter, with the same characteristics, was installed to overcome possible failures of the first one without any delay.

A pre-existing 10 kVA diesel generator (380/220 V), is used now as back-up equipment and can operate in two different ways (also simultaneously):

- feeding directly the A.C. loads (in particular a 380 V 3Ø dish washing machine)
- recharging batteries by means of a rectifier installed in the same cabinet with the inverter.

The diesel generator can be operated only by manual commands.

4. DETAILS OF P.V. USER APPLIANCES

As mentioned in the introduction the three country-houses of "Prato Rondanino" centre are differently inhabited during the year according to the season; the relevant A.C. loads are listed in the following tables:

A) Resident houses

Load	Unit Power	Number
Refrigerator	200 W	2
TV - Set	50 W	2
Fluorescent Lamp (1200 lumen)	25 W	20

B) Tourist - House

Load	Unit Power	Number
Refrigerator	650 W	1
TV - Set	50 W	2
Fluorescent Lamp (1200 lumen)	25 W	20
Fluorescent Lamp (900 lumen)	15 W	20

Three different situations can be distinguished during the year:

- from November till February, the Tourist-House is closed, so only resident users have to be supplied;
- tourist concourse and consequent energy requirements reach a maximum in June, July and August;
- in the other months, a peak energy consumption occurs during weekends.

Other kitchen appliances (slicer, mixer ....), negligible in respect to energy consumption evaluation, are used. A 380 V - 3Ø dish-washing machine can be fed only by the diesel generator.

Detailed information on the plant management, and energy balance will be available as soon as the data acquisition system will be put into service.

5. MONITORING SCHEME

An automatic data acquisition system will monitor the following parameters, according to the suggestions received with the document "Guidelines for Photovoltaic - system monitoring" both from "Joint Research Centre" and from the "Directorate General of Energy" of the "Commission of European Communities".

Parameter	Symbol	Units	Accuracy
Global solar radiation (horizontal)	IG	kWh/m^2	Calibrated cell
Global solar radiation (array plane)	IA	kWh/m^2	" "
Max. wind speed	VM	m/s	± 2 m/s
Wind direction	VD	degr.	± 10 degr.
Max. ambient temperature	TA	degr. C	± 2 deg. C
Max. module temperature	TC	degr. C	± 2 deg. C
Array - output energy	EA	kWh	± 2%
Energy to (+) and from (-) battery	EB	kWh	± 2%
Inverter/input DC energy	EII	kWh	± 2%
Inverter/output AC energy	EIO	kWh	± 2%
DC energy from auxiliary diesel generator (rectifier output)	EG1	kWh	± 2%
AC energy from auxiliary diesel generator	EG2	kWh	± 2%

The energy to all AC loads (ELAC in the energy flow diagram-fig.1) equals EIO or EG2 (the inverter and the diesel generator can't run in parallel).

All numerical data will be hourly averages, or maximum values, of measurements taken at one minute intervals. They will be recorded on ECMA 46 magnetic tape cassettes in the format suggested in the above-mentioned document.

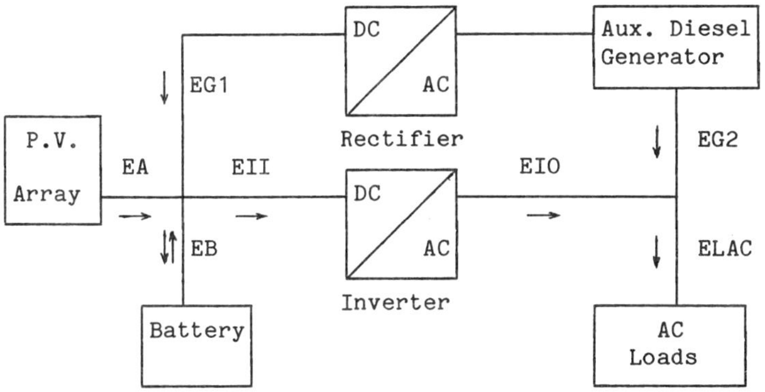

Fig. 1 - Energy flow diagram

6. EXPERIENCE OBTAINED DURING INSTALLATION

No. problem encountered as regards the P.V. system.

7. ESPERIENCE OBTAINED DURING OPERATION

Some minor troubles occurred in the DC-AC converters due to the ambient conditions (condensate, diesel exaust gas, etc.). The tropicalization of the main components, such as electronic cards and power transistor sections, have definitely solved the problem.

CONTRACT NR SE 713/84

DEMONSTRATION PROJECTS FOR STAND-ALONE SYSTEMS WITH RENEWABLE ENERGY SOURCES

H.K.KOETHE and E.ROESSLER
Fraunhofer Institut für
Solare Energiesysteme (ISE)
Oltmannsstr. 22
D - 7800 Freiburg

Summary

The Demonstration-Project consists of a subproject A "Information Pavillon" and a subproject B "Water Pumping System". Subproject A has been postponed to 1987 due to budgeting reasons. Subproject B was taken into service in April 1986 for the opening of the "State Garden Show, Freiburg, 1986".

The water pumping system consists of a PV-array, a lead acid battery, two DC-water pumps and a wind generator as a back-up. It raises up to 2 l/s ($7.2 m^3/h$) to a height of about 14 m and would be able to supply a village in Africa with approx. 5000 people with drinking water.

After taking into use, the main remaining problems were the development of a start-/stop-circuitry for the water pumps, as well as a redesign of the elements and wiring of the electronics cabinet and the initiation of the data acquisition system.

All these problems have been solved and the data acquisition period had started in October 1986. The continous overall efficiency of the water pumping system could be demonstrated to be approx. 5 %.

1. INTRODUCTION

The water pumping system is located on the area of the "State Garden Show" in Freiburg with a latitude 48.0°N and altitude 7.84°E. It consists of a PV-array with a peak-power of 1.73 kWp, a lead acid battery with 350 Ah and a nominal voltage of 64 V, two DC-driven water pumps with a power of 0.4 kW and a wind generator with a nominal power of 0.6 kW at a wind speed of 8 m/s.

The water pumping system raises a water quantity of 2 l/s (7.2 m^3/h) to the top of a hill with a height of about 14 m for ten hours each day. The two pumps are mechanically connected in series, so that one pump feeds the other, in order to reach the needed height.

In spring and in autumn, the wind generator supports the solar generator with the power supply. The water pumps are controlled by a special start-/stop-control logic, which ensures that the current for the water pumps is less than 6 A at all times. The whole system is controlled by a control-logic, which cuts off the water pumps if the battery experiences a deep discharge or the system fails.

2. PHOTOVOLTAIC ARRAY

The solar array is orientated towards the south with an inclination of 45°, and is mounted with a special supporting structure on a steel grid tower. It consists of 45 solar modules, type PQ 10/40/01, manufactured by AEG, with a total peak power of 1.73 kW (STC).

Because we did not use a MPPT (maximum power point tracker), it was necessary to connect 5 modules in series to get a nominal voltage of 64 V. Consequently we have 9 sub-arrays, which are connected in a switch box on the top of the tower. This switch box also includes the protection against lightning and overvoltages.

3. POWER CONDITIONING

For the water pumping system, a lead acid battery, consisting of 64 battery cells with a capacity of 175 Ah, type Berga 4BB27 is used. The storage battery is connected in two groups to get a nominal voltage of 64 V. The advantage of using a storage battery in a water pumping system is to use low currents of the solar generator too, which normally could not power the pumps, if they were connected to the solar generator directly.

The system will be mainly controlled by the state of charge of the storage battery. In normal operation the solar generator and, if there is a sufficient wind speed, the wind generator too provide the battery with DC-energy.
A shunt regulator will prevent the battery from being overcharged by limiting the charge-voltage to a level of 2.35 Volts per cell. The battery provides energy directly to the two water pumps which are controlled by a timer also. If the battery is deeply discharged or any other failure happens to the system, the control-logic will switch off the water pumps. The plant only returns to work, if the battery voltage reaches 2.2 Volt per cell on charging or the fault is remedied off manually.

For back-up power supply, a wind generator mounted on the top of the tower is used. Its cut-in windspeed is at 3 m/s and it reaches the nominal power of 0.6 KW at a windspeed of 8 m/s.

4. DETAILS OF PV USER APPLIANCES

The chosen 60 VDC/400 W - centrifugal pumps type AQUASOL 50L manufactured by KSB are normally intended for direct connection to solar modules with a limited short circuit current (I_{sc} ca. 7 A max.). Because the pumps are not allowed to be operated from batteries, a special starting circuitry had to be developed in order to limit the motor current to safe values (less than 6.0 A) under all circumstances (e.g. starting of the pumps, pumps blocks mechanically etc.).

5. MONITORING SCHEME

At the water pumping system, the following parameters are monitored:

5.1 meteorological data

- insolation (array plane) IA
- wind velocity VM
- wind direction VD
- solar module temperature TC

5.2 DC-energies

- solar generator current and voltage
- shunt current
- current to DC-loads
- voltage DC-loads
- wind generator current and voltage
- battery voltage

All recording is performed by a data logger type Modas 1223 manufactured by NES, which has 16 analog input channels and 7 pulse input channels with 12 bit resolution. Readings will be taken every 2 seconds and average values calculated from these readings will be recorded every hour.
The data logger is able to take up to four plug-in memories (RAM), which have an on-board lithium battery. An advantage of this data logger is that it has a very low power consumption of 0.5 W and a sealed on-board lead acid battery. Reading the plug-in memories and calculating the energy flow is done by an IBM-PC using a special equipment, which transfers the data from the memories via an RS 232 interface to the IBM-PC.

6. EXPERIENCE OBTAINED DURING INSTALLATION AND COMMISSIONING

The water pumping system has been put into operation on April 1986 along with the "State Garden Show in Freiburg 1986". Although some provisional installations were made, the system operated properly from the start. According to the original time schedule the Data Acquisition Unit (DAU) should already be in service. Instead, the solar generator, wind energy converter, batteries and pumps are measured by using Ah-integrators. The DAU was put into operation after installation of the electronic start-/stop circuitry in September 1986.

7. EXPERIENCE OBTAINED DURING OPERATION

The development of the start-/stop circuitry for the pumps proved to be more difficult than previously expected. The main reason is the fact that the pumps not only had to be protected against faults in the water circuit, but also against failure of components in the electronic circuitry. These problems have already been solved.

Due to the long distance of about 50 m between the solar generator and the shunt transistors, protecting the battery against overcharge, high voltage transients across the transistors occured on current turn-off, which consequently can damage the transistors. One of the tasks of the following period therefore will be the solving of this serious problem.

First measuring results of the data acquisition from October 1986 have been sent to JRC-ISPRA. The analysis lead to the following results:

MONTHLY ENERGY BALANCE
(KWh)

Total incident solar energy (array plane)	2351.1
Total array output cncrgy	136.7
Total energy I/O (+/-) to/from batteries	53.7
Useful PV energy production	83.0

FIGURES OF MERIT (%)

Array efficiency (max.)	6.32
Array efficiency (mean)	5.81
Array eff.(mean) in % of nominal eff.	76.27
Intermediate efficiency	75.24
Cell Temperature (°C) (mean)	20.84
Cell Temperature (°C) (max.)	49.50

CONTRAT N° SE 696 84 FR

PHOTOVOLTAIC EQUIPMENT FOR 12 LODGINGS AND REFUGES LOCATED IN THE FRENCH ALPS

GRANDE TRAVERSEE DES ALPES - 14 rue de la République 3800 GRENOBLE
ASSOCIATION POUR LA PROMOTION DES ENERGIES NOUVELLES - Le Boulou
38410 ST MARTIN D'URIAGE

Summary

Electrification, by means of photovoltaic solars cells, of 12 lodgings and huts in mountains, regularly scattered along the great crossing of the French Alps, from the lake of Geneva to the Mediterranean, in order to develop the practice of hiking and trekking by improving welcome of hikers and life conditions of keepers.

The altitude of the concerned lodgings is varying from 1500 m to 3000 m. This electrifying has several objectives :

- to use an energy source avoiding transport of fuel
- to adopt lodging to the sociological evolution of practices of sport life in a mountain environment.

The principal phases of the project are as follows :

1 - dimensioning - technical study
2 - realization (fabrication, assembling, wiring)
3 - transport, installation, essays and setting up
4 - measurements

The whole project can be considered as innovating.

Actually, we are in the end of the last part of the project : MEASURING

- seven refuges are already coming in the state of measuring since July 1986
- the measuring installation for five refuges are waiting, we had planned their measuring installation during the spring of 1987.

1. INTRODUCTION

The innovatory aspects of the project consist essentially in the mode of use of the Photovoltaïc energy.

From a technical point of view they are :

- to test solar plants in severe climatic conditions on a large scale and therefore to prepare a next stage of diffusion.
- to obtain a renewable source of energy and then to avoid regueling of classical plants.

From a sociological point of view :

- to allow the ramblers to have a lodging corresponding to the evolution of present needs.
- to break off lodging keepers' loneliness, and improve working conditions.

With an average time of inhabitation of lodging, 1 kWc' effective produces 700 KWh/year. Taking an average power of 400 Wc' per lodging, the annual production is : 3360 KWh/year.

The total effective power is : 5.3 kWc'.

It is difficult to compare this project to a conventional installation. The two possible solutions, an electric line connected with the national network or a fuel powered electric generating set, are incompatible with the respect of mountain environment.

2. TECHNICAL PRESENTATION

2.1. Rating

In order to simplify assembly and installation, lodgings were classified in 3 types corresponding to following peak power :

- A type : 280 Watts peak
- B type : 480 Watts peak
- C type : 600 Watts peak

The estimation of using time of the PV plant gives the following results for dimensioning the elements of solar plants :

	A TYPE	B TYPE	C TYPE
Useful PEAK POWER	280	480	600
Number of panels PWP 40 PHOTOWATT	7	12	15
Summer daily consumption	815 Wh	1050 Wh	1300 Wh
Winter daily consumption	826 Wh	1184 Wh	1309 Wh
Number of batteries 12 V 105 Ah	7	12	15
Stored energy	8820 Wh	15120 Wh	18900 Wh

Stored energy permits several working days without sun.

For reliability reasons and taking into account maintenance problems photovoltaic plants were divided in independent modules according to the following scheme. The originality is in the choice of several regulators :

- For A type : 3 Photowatt R10C12
- For B type : 4 Photowatt R10C12
- For C type : 5 Photowatt R10C12

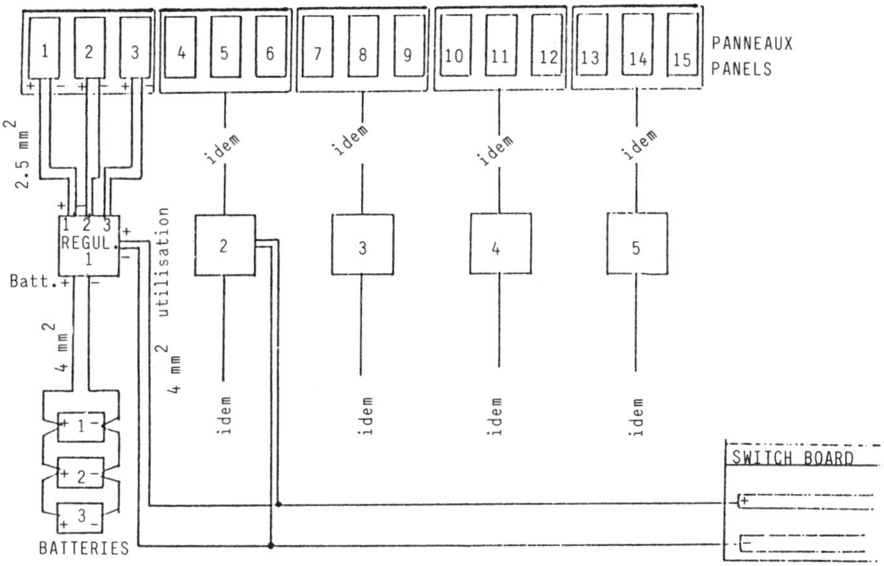

Ex : 15 panels distributed in 5 independent plants with 3 panels

2.2. Individual technical study for each lodging

The study of each lodging begins with a complete visit in order to verify the orientation of the frontages, the identification of materials used in building, the evaluation of masks and a first meeting with keepers. After this verification, we established :

- panel's installation diagram
- interior wiring diagram
- repartition panel diagram

South Front

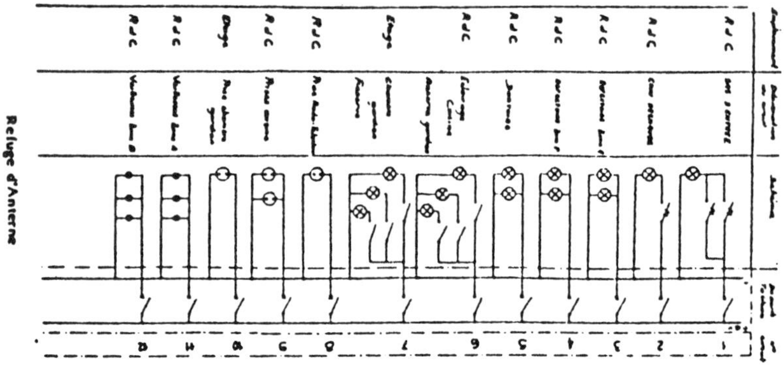

SWITCH BOARD

Remark : Some of the lodgings are not kept during the winter season. They remain open to users. In this case, 2 possibilities were given
- to turn off the electric installation
- to equip the installation with electronic timer driven by a push-button.

This solution delivers electricity during a given time : 5 minutes in bed rooms and 30 minutes in refectory.

2.3. Fabrication - Assembly
2.3.1. Frames and Fixing

Photovaltaic PWP 402 panels are fixed on metallic frames treated against corrosion. Each frame can support 3 or 4 panels.

1) cadre : cornière acier
40 épaisseur 4

2) entretoise : plat acier
60 épaisseur 4

3) patte anti-vol
Plat acier 40 épaisseur 4

4) patte de fixation
Plat acier 40 épaisseur 4

NB : Traitement anti-corrosif
- métallisation
- wash-primer
- peinture

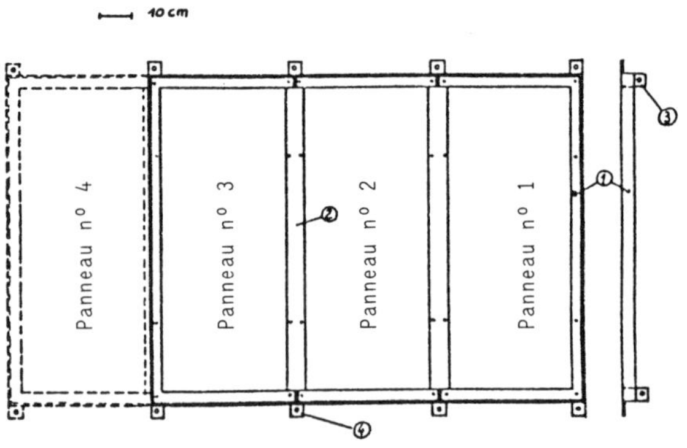

METALLIC FRAMES

Frames were fixed :

- on frontages (BOSSETAN, ANTERNE, EMPARIS, LONGON, NICE, SOREILLER)
- on the roof (LEISSE, CERVIERES, VALBELLE)
- on external supports (LA PARE, CHARDONNET, FONDS DES FOURS).

Each time it will be necessary and possible, two tilting angles were provided : one for summer, the other for winter.

2.3.2. Battery box

For lodgings which are particularly exposed to low temperatures batteries are set in a thermally insulated box. An electrical resistance (4 W, 12 V) driven by thermostat maintains the temperature inside the box between convenient values. The bottom of the battery box has a retention can.

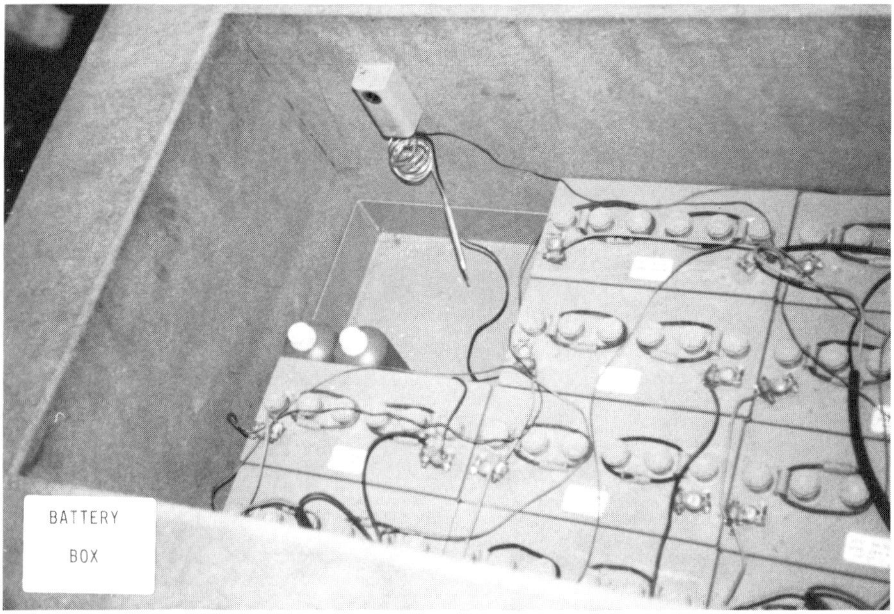

3. TRANSPORT, SETTING UP AND ESSAYS

Transport was made either by helicopter, by 4WD vehicles or by human means.

The PV plants were set up at following dates :

Lodging name	Setting up date
BOSSETAN	07/85
ANTERNE	07/85
EMPARIS	09/85
LA PARE	09/85
LONGON	07/85
NICE	07/85
FONDS DES FOURS	07/86
LA LEISSE	07/86
FONDS DE CERVIERES	10/86
SOREILLER	07/86
CHARDONNET	09/86
VALBELLE	04/87

All plants are systematically verified by A.P.R.E.N. which give a conformity certificate.

4. MEASUREMENTS

4.1. Measurement equipment

4.1.1. Energy meter

In accordance with the prescription contract, lodgings are equipped with energy meters. They measure during 2 years :

- Energy displayed by PV panels
- Energy consumed by users

Meters are plugged as indicated in the following diagram. We have chosen Wattheuresmetres commercialized by SCHLUMBERGER.

Displayed energy is measured on one PV panel for lodgings having one orientation and on one PV panel of each orientation for lodgings having several orientations. The resulting displayed energy is the sum of the displayed energy of each PV panel multiplied by the number of panels of each orientation.

The regulating system is fractionned and it is not possible to have exactly the same level on each regulator. The result of that system is an uncertainty of the displayed energy.

Daily, the lodging keeper notes the meter values.

4.1.2. Detailed measurements on 2 lodgings

A.P.R.E.N. obtained near A.F.M.E. a contract to make detailed measurements on 2 lodgings. A schematic lay-out is given in the following figure.

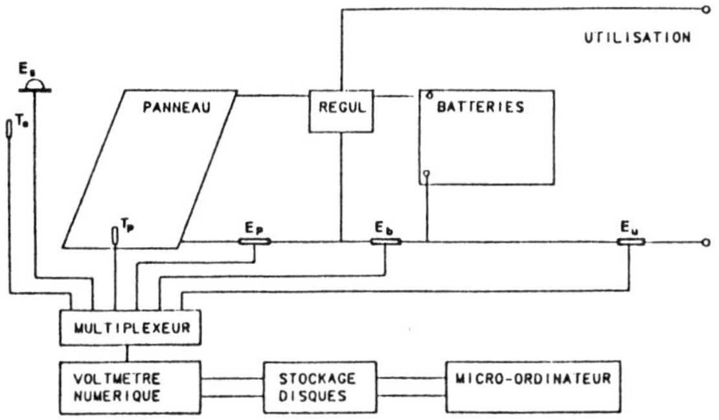

SCHEMA DE PRINCIPE

The measurement station is powered with 12 V DC delivered by the PV plant and all devices are placed in a hand-bag thermally insulated and conditioned. Insolation measurements are made each 10 minutes in accordance with CEE recommendations. The data recording frequency of other parameters is 5 minutes with integration during one hour. Data are stored on discs and sent to A.P.R.E.N. every 3 months for calculations and results. The following parameters are monitored by the station :

- Insolation received on panels
- PV panel temperature
- Ambient temperature
- Produced energy
- Energy received or delivered by batteries
- Consumed energy

4.2. Results

4.2.1. Results from energy meters for 1986

The following table gives the results obtained from lodgings which are presently equipped with energy meters.

Lodging	Panel numb.	Counting period	Total consumed energy Wh	Daily mean consumption Wh	Daily estimated consumption Wh	Daily panel energy Wh
BOSSETAN	15	23/06-03/10	35970	353	1300	23.5
ANTERNE	7	04/07-28/09	19610	228	826	32.6
SOREILLER	15	01/08-13/09	24880	592	1300	39.6
LA PARE	7	25/08-14/09	5860	293	826	41.8
NICE	15	27/05-10/11	199470	1194	1300	79.6
EMPARIS	12	01/08-01/10	13270	218	1050	18.2

These results show us large differences in the way of using energy between lodging keepers. Some of them economize much the energy. In all cases the consumed energy is lower than predicted but all lodgings are not fully equipped with electric accessories and it seems that keepers dare not use too much energy because it's a new technique for them.
Daily consumption example :

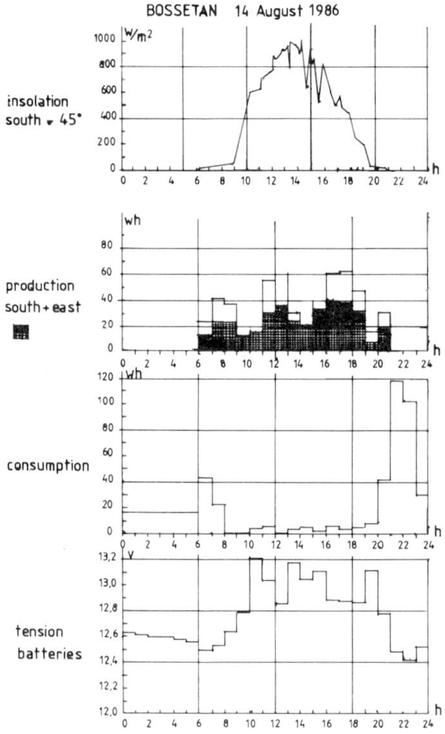

4.2.2. Detailed measurement

Since 07/86, BOSSETAN is equipped with an automatic measurement station. It is possible to see an example of daily results obtained with this station in the following figure.

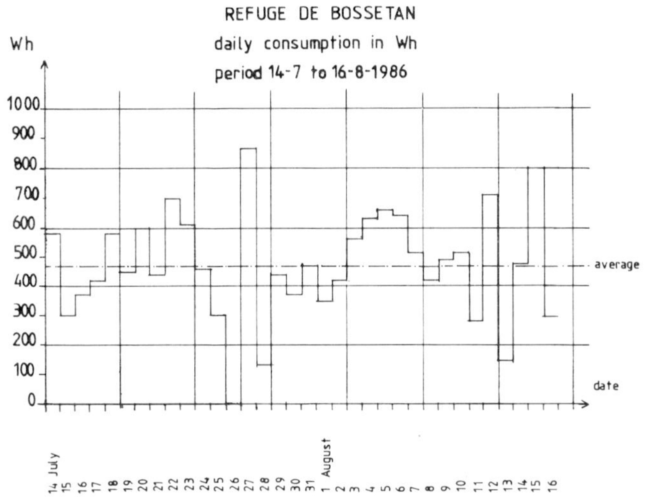

5. PROBLEMS

5.1. Setting-up problems

The companies which realized the plants have no previous specific ability with PV systems. Low voltage DC currents seem to give them some problems in the choice of electrical components and A.P.R.E.N. must be very careful about plants and does not hesitate to refuse it if some elements were not in conformity with projects. But after this first experience, all companies enjoy to experiment this new technique.

5.2. Using problems

In spite of recommendations, some keepers used non designed devices on the plant causing some trouble and occasionally short circuited the PV plant.

The first series of lights used give no satisfaction and after only some using hours many lights were out. Now they are changed with other kinds of lights.

5.3. Measurement problems

We have two types of problems with measurements :

- Technical problems with meters in measuring low levels of consumption
- Human problems with keepers which forget to note daily the measurements on the meters.

6. CONFERENCES AND PRESS

6.1. Conferences

This demonstration project gives rise to communication in several conferences and colloquiums :

- "Colloque sur la maîtrise de l'énergie en montagne", 30/31 mai 1985 TIGNES (FRANCE) par A. BONTEMPS, C. GOUBIER, C. MARQUET, J.C. SOLECKI (A.P.R.E.N.) et J.P. LYARD (G.T.A.)

6.2. Press

Press and Medias were represented at the inauguration of the plants and some papers were published on this occasion.

One article was published about this realisation in "Cahiers du batiment" (02/1986).

CONTRACT NR SE/313/83 HE

GAVDOS PHOTOVOLTAIC POWER PLANT

J. CHADJIVASSILIADIS * and H. KADELLA **

* Public Power Corporation, Athens/Greece
** Siemens Solar GmbH, Bergisch Gladbach/FRG

Summary

Gavdos is a remote island located in the libyan sea, south of Crete, and it is the southest part of Europe. No electricity is existing on the island, but it offers the most favorable conditions for solar radiation and photovoltaic applications. The aim of this project is to demonstrate photovoltaic generators as technically suitable, economically achievable and socially well accepted today solutions to meet the electrical energy requirements of remote islands. The concept for the electrification of the island is to install a central photovoltaic power station to supply the existing small communities (phase II), while despersed houses will be electrified by small independent photovoltaic units under another project (phase I). During Summer of 1986 an autonomous 20 kWp photovoltaic Power Plant is installed to cover the electrical needs of the inhabitants in the two communities of the island. Seasonal adjustment of the tilt angle is provided in order to adapt the output of the PV generator to the demand. Two transmission lines leading from the PV-plant to the two communities and the relevant network have been erected to supply the consumers with AC 220/380 V, 50 Hz.

1. INTRODUCTION

Gavdos is a remote island located in the Libyan Sea, south of Crete and it is the southest part of Europe with a Latitude 34° 50' N and Altitude 24° 05' E. Two communities are located on the island Kastri and Vatsiana and some dispersed houses. The population varies between 70 persons in Winter and 90 persons in Summer with additional visitors during the Summer season. A central photovoltaic plant of 20 kWp is installed between the two communities and two lines for three phase A.C. will supply the consumers, one for each community (see fig. 1). An area of 3000 m^2 was necessary for the installation of the plant. An architectural design was elaborated adapting the plant to the environment. An old windmill is used for the installation of the electronical equipment and a small housing for the battery storage was built in a traditional way.

2. PHOTOVOLTAIC ARRAY

The rated power of the photovoltaic arrays is 20 kWp. 160 monocrystaline silicon modules type SM 144 manufactured by Siemens AG are installed in 8 arrays (see fig. 2). The dimensions of the modules is 1020X1470 mm mounted 144 round cells of 10 cm diameter with a rated power of 130 Wp. The module frame is of anodized extruded aluminium, the module thus being capable of withstanding high surface-loading. The module is equipped with protection diodes which are located within the terminal box. 20 modules are connected in series to one branch and mounted on one supporting structure (see fig. 3). The supporting structures are simply designed by using H section and tubes of hot galvanized steel. A good stability of the supporting is achieved by placing the middle of the module above the upper mounting tube. The construction provides an easy seasonal adjustment of the tilt angle. Lightning conductors and earthing system is installed in the generator field to protect the installation. Special devices by varistors are used also to protect the modules and the electronic equipment against lightnings.

3. POWER CONDITIONING

A storage battery consisting of 250 lead acid elements (2 V, 468 Ah each one) connected in two parallel groups (250V, 936 Ah) is used, manufactured by TUDOR HELLAS.

A new type of solar inverter is developed for this project. In one cubicle are mounted both the DC/DC converter and the DC/AC inverter, (see fig. 4).

The task of the DC/DC part of the solar inverter is the decoupling of the photovoltaic generator voltage from the battery voltage. In this way the solar generator can be utilized in an optimal way. In addition an MPP-Tracker is conected to utilize the solar field optimally. The efficiency of the DC/DC part is extremely high (96%).

The DC/AC part is of 30 kVA rated power, three phases 230/400 V, 50 Hz. The efficiency is very high, 94% at the rated power and above 90% at 20% of the rated power. The no load losses are limited to 800 W. This inverter is designed to operate in four quadrants for a possible integration of a wind converter when the demand will be increased in the future. The solar inverter is installed in an old windmill while the storage battery in a small housing.

4. PV USER APPLIANCES

The consumer will be provided by low energy consumption domestic appliances such as lighting lamps, TV sets and refrigerators. In additional special lamps will be used for street lighting :

- for the illumination, special energy saving lamps 18 W, 900 lm
- for refrigerators, 120 l with very good isolation
- TV sets, colour TV 50W with 16" tube.

The capacity of the plant can meet the demand of the consumers during summer and winter by a seasonal adjustment of the modules tilt angle.

The transmission lines and the distribution system are designed to have very low losses and to protect the plant against short circuits. A superimposed control is provided to receive data and to act to the solar inverter and switching devices.

5. MONITORING SYSTEM

A monitoring system developed by INTERATOM according to the CEC/ISPRA quidelines is installed for the beginning of operation of the plant, for a period up to the time when the monitoring system provided by PPC will be ready for installation. This data aquisition system is also designed according to the ISPRA quidelines with equipment of Hewlett-Packard.

The H-P 3421A Data Aquisition unit will be used with a capacity of 30 analog and/or digital indication inputs. The HP71B Handheld computer is connected to the Data Aquisition unit via a HP-IL link. The HP-71B computer includes a quartz crystal clock and can turn on the HP3421A unit, trigger to scan a list of up to 30 inputs, instruct a mass memory to store the readings and then power down the entire system until the next time passes. The HP-9114B portable Microfloppy shall be used as mass storage media.

The active system shall be supplied by 12VDC. The peak power consumption is estimated to be less than 200 W with an average power consumption less than 25 W.

The data scanning of the required measurements will be done on a periodic basis with a time interval of the order of about 10 seconds.

By an appropriate subroutine running in the HP-71B computer the average values of the sampled measurements will be calculated and stored in the computer's RAM every 10 minutes. The 10 minute averaged values will be transfered from the computer's RAM to the floppydisc every hour. The floppydisc will be transferred in Athens every month and from a central system a copy of a floppydisc will be produced and sent to

CEC/ISPRA. The analysis of data for several useful conclusions will be done in Athens in a small central system (H-P86B).

6. INSTALLATION AND COMMISSIONING

An access road to the site has been constructed and the area prepared for installation. A small housing for the batteries is built and the old windwill is restored for the installation of the electronic equipment.

All the components of the plant have been transported from Germany and Athens via Crete to the island. The transportation from south Crete to the site was a very difficult task due to the lack of infrastructure and site supporting devices.

The plant has been erected during the summer season 1986 and it was completed at the end of October. Because of the difficulties to reach the island in the winter season the commissioning was forseen for the first months of 1987. The commissioning started the first days of March 1987 but unfortunately it had to be interrupted due to the extremely hard weather conditions unusually for this region. Therefore the commissioning will be continued on the first days of May.

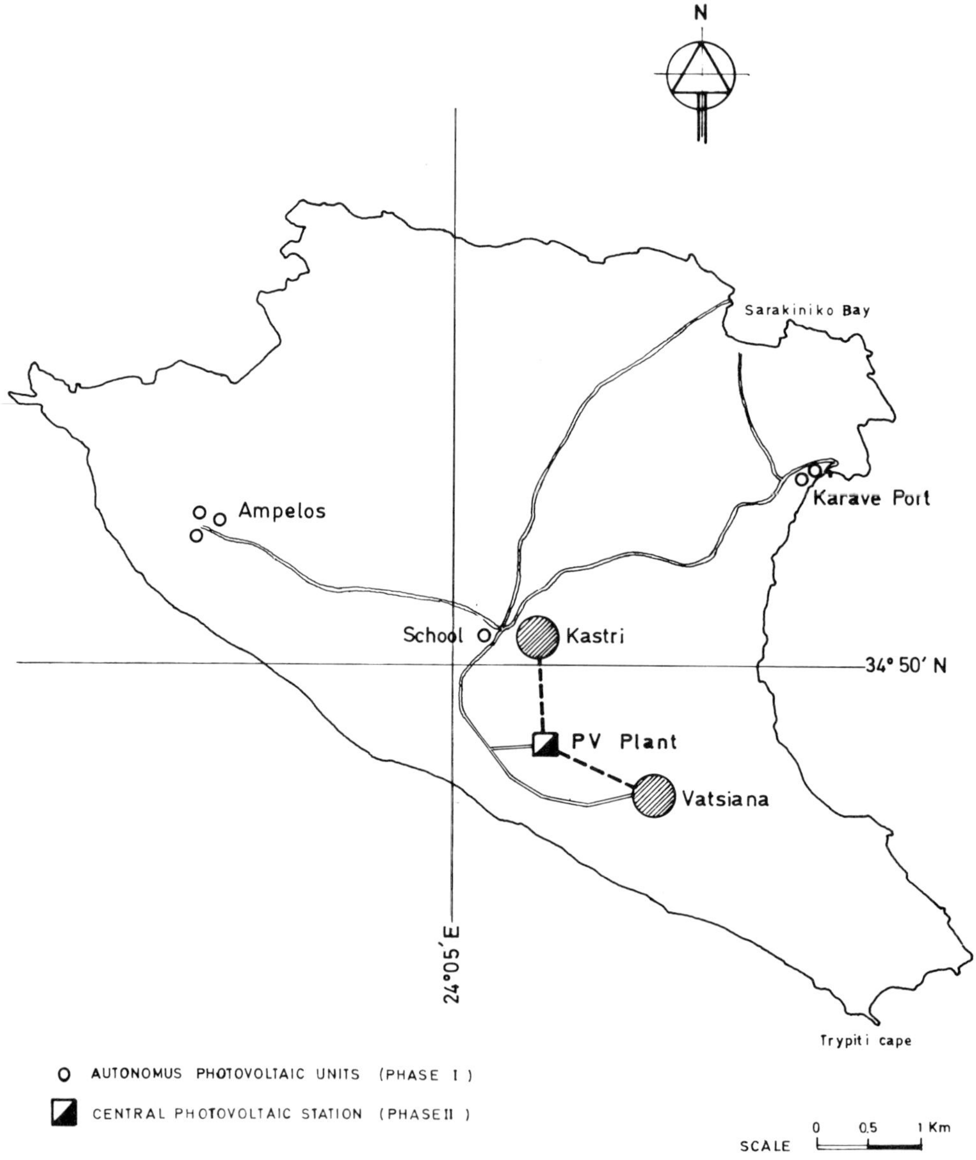

Fig. 1 GAVDOS ISLAND
THE CENTRAL 20 KWp PV POWER STATION

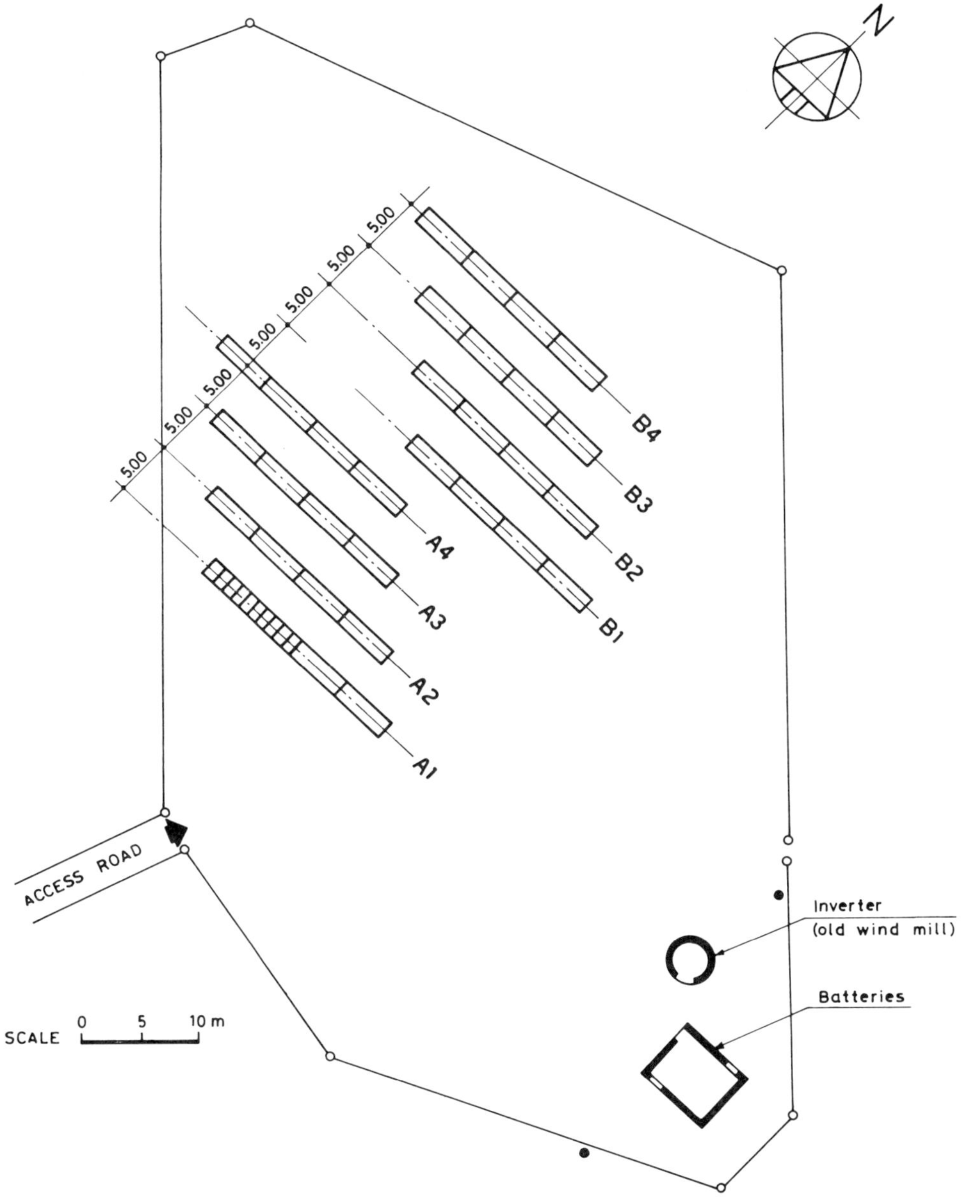

Fig. 2 GAVDOS PV PLANT 20 KWp
GENERAL LAYOUT

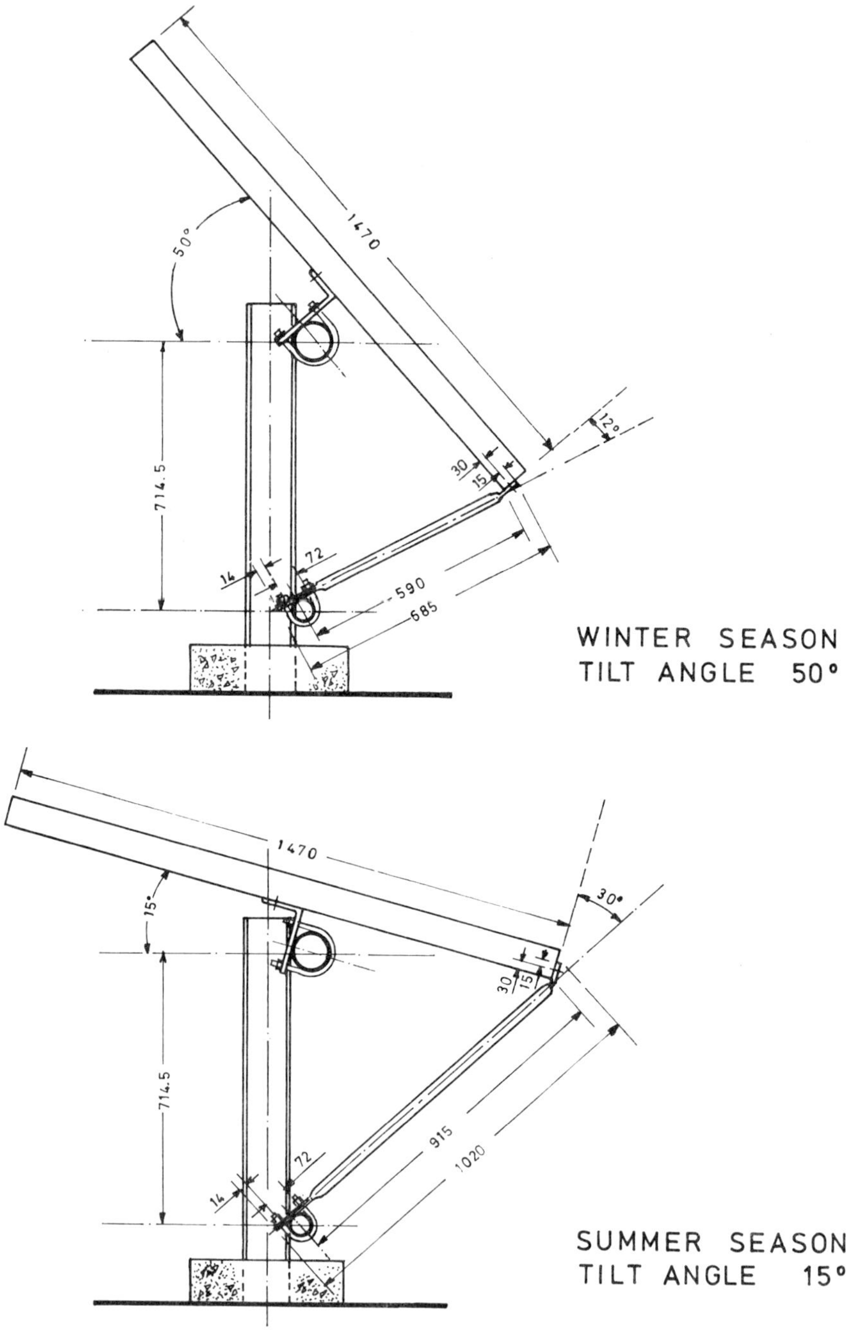

Fig. 3 GAVDOS PV PLANT 20 KWp
PANNEL SUPPORTING STRUCTURE

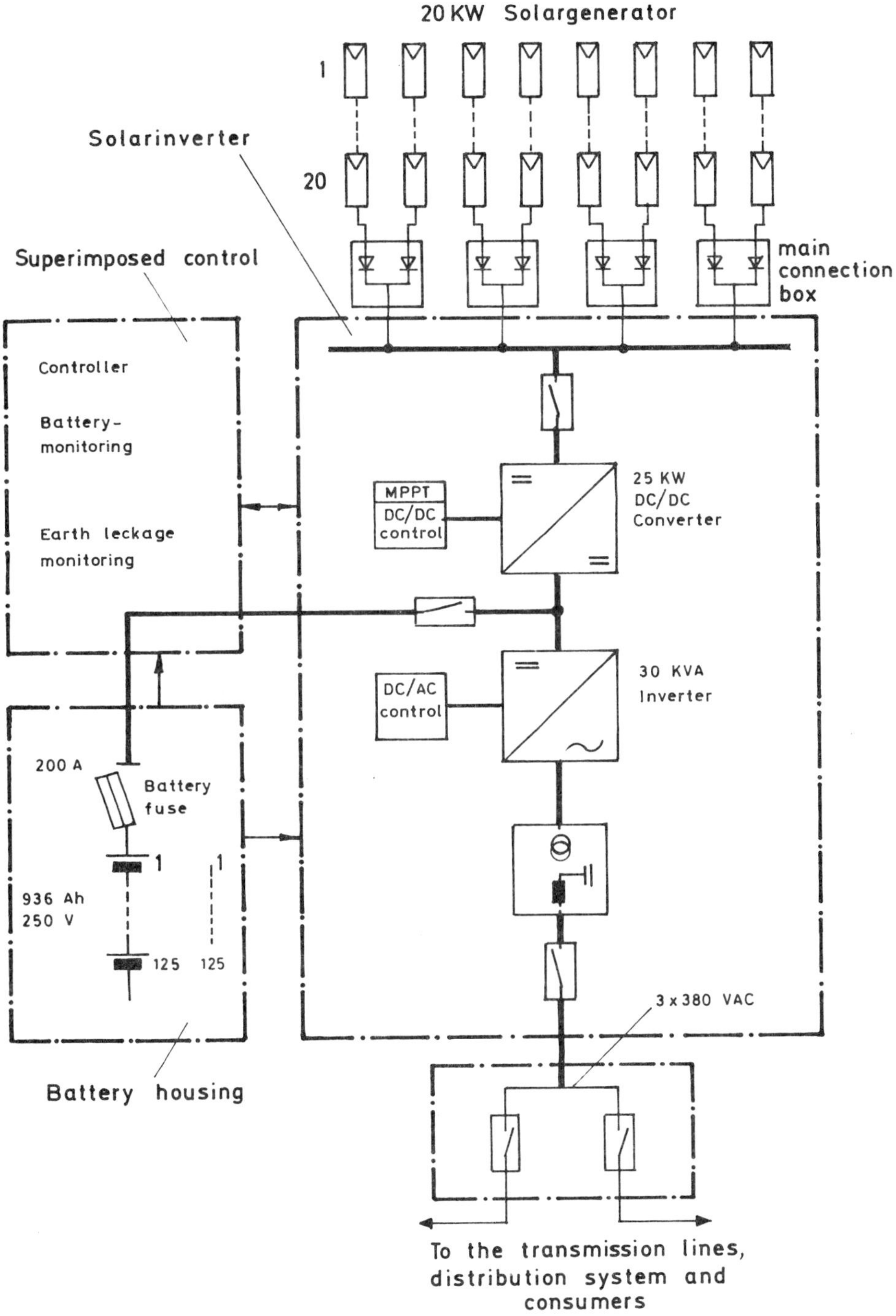

Fig. 4 GAVDOS PV PLANT 20 KWp
BLOCK DIAGRAM

CONTRACT NR SE/314/83

ANTIKYTHIRA SOLAR PHOTOVOLTAIC POWER PLANT

J. CHADJIVASSILIADIS * and H.J. LOWALT **

* Public Power Corporation, Greece
** AEG Aktiengesellschaft - F.R.G.

Summary

Antikythira is a small island located between Kythira and Crete. Several small communities and houses are dispersed in the island. Solar photovoltaic generators will be used for the electrification of the island except the harbour Potamos which recently has been electrified by a diesel power station. The aim of this project is to demonstrate photovoltaic applications as technically suitable, economically achievable and socially well adapted today solutions to meet electrical energy needs of dispersed communities and houses in a small island. A central photovoltaic power station of a capacity of 25 kWp will supply the nearest communities and twelve small autonomous photovoltaic units will supply the remote houses. Low energy consumption domestic appliances will be used by the consumers. Lead acid storage batteryisused to supply continuously the consumers day and night, even in cloudy days.
The plant is designed to operate automatically without attendance, to cover the demand in low season (winter) and high season (summer). The central plant is in a trial operation from April 1987 supplying the street lights. After some additional tests, the central plant and the autonomous PV units will be ready to supply the houses.

1. INTRODUCTION

Antikythira is a small island located between Kythira and Crete, its latitude is 35° 52' N and altitude 23° 18'. One community Potamos is sited in the port and some others are dispersed in the middle of the island. In addition some houses are existing in other sites of the island.

A diesel power station has been recently installed to supply the community of Potamos. The electrification of the other communities and remote houses is provided by solar photovoltaic power generation under this project. A central photovoltaic power plant of a 25 kWp capacity and 12 small autonomous PV units of 0,7 kWp each one for 12 remote houses are provided.

The central power plant fed the consumers by 220 V/380V A.C. while the small independent units by 24 V D.C.
Low energy consumption domestic appliances will be used by the consumers.
The population varies from 70 inhabitants in winter season to about 120 in summer season.

2. PHOTOVOLTAIC ARRAY

The rated power of photovoltaic arrays is 25 kWp for the central PV power station and 8,4 kWp for the 12 remote houses. The cells are of polycrystalline silicon manufactured by AEG.

The module consists of 40 cells connected in series. All interconnectors are welded. The cells are embedded like a sandwich between two hardened class sheets and plastic layers. For an optimal way of module installation and to reduce the structure costs, each module is equipped with mounting fittings. The framing of the module is done with stainless steel so a very long lifespan is promised.

The electrical data of the modules may be taken from the following data sheet.

Characteristic values (AM 1.5 – 100 mW/cm2)	Solar Cell Temperature 0 °C	25 °C	60 °C
Open circuit voltage (V)	24.6	22.4	19.2
Short circuit current (A)	2.37	2.41	2.46
Current at max. power (A)	2.18	2.20	2.23
Nominal power (W)	42.6	38.4	32.4

16 Modules are connected in series to form a string. 45 strings perform the whole generator. The generator devided into 8 switchable groups which are connected via underground cables to the DC-busbar. Each group is equipped in field with varistors for lightning protection. In addition lightning protection spikes are installed.

The supporting structures are simply designed by using normal profile sections of hot galvanized steel. The tilt angle of the module is constant and was find 43° as the optimum angle to meet the demand for both summer and winter seasons.

3. POWER CONDITIONS

The inverter is a 30 kVA self commutated inverter manufactured by AEG for the stand alone operation mode, built up with fully controlled thyristor bridges and a comparable output quality with utility boards. For the operation in remote areas it is of great importance to assure all safety preparations for the human life and to guarantee a reliable function.

To realize this well adapted, function of the whole system an industrial type of inverter is used.

Technical Data

Self Commutated Inverter

Nominal Power AC	:	30 kVA
Voltage AC	:	400 / 230 V 50 Hz
Voltage deviation range	:	+/- 2% at symetrical load
Frequency deviation range	:	+/- 0,5%
Unsymetrically load (possible)	:	100% (neutral line carry full load)
Efficiency	:	91% (100% load,load factor 0,8)
	:	84% (20% load, load factor 0,8)
Harmonic distorsion	:	5 %
Voltage variation range DC	:	+/- 20%
Nominal voltage DC	:	220 V
Overload capacity	:	120% for 10 min
Short circuit	:	proof
Voltage AC adjustable	:	+/- 5%

4. BATTERY STORAGE

The storage battery consists of 216 lead acid elements of 546 Ah each. The elements are connected in two parallel groups consisting a storage system of 216 V nominal voltage and 1092 Ah nominal capacity.

The batteries for the single houses is of 24 V 300 Ah for each one consisting of 4 cells.

5. INSTALLATION OF THE PLANT

For the installation of the central power plant an area of 3000 m^2 has been used located in the center between the communities. The arrays have been arranged as in the figure is shown. In the north part a small housing has been erected for the installation of the batteries and the electronic equipments.

The foundation of the supporting structures of the modules is made by concrete.

An extended distribution system has been erected to supply the houses, (Fig. 3).
All the materials for the installation has been transported by ship from Pireus and South Peloponnesus to the port of Antikythira and then by a tractor and carriage to the site.

The modules, the electronic equipments and the refrigerators are transported by lorries from Germany to Pireus and then by the same way to the site.

6. PV USER APPLIANCES

Low energy consumption appliances will be used by the consumers.
- for illumination, special lamps of 18 W, 900 lm for the houses and street lights
- refrigerators of a capacity of 110 l and 70 W motors with very good isolation
- TV sets color TV of 50 W and 16" tube

For the single houses the same appliances are provided in 24 V DC.

7. MONITORING SYSTEM

A data aquisition system according to the quidelines of Ispra has been designed with data aquisition and instrumentation equipment of Hewlett - Packard.

The H-P 3421 A Data Aquisition unit will be used with a capacity of 30 analog and/or digital indication inputs. The H-P71B Handheld computer is connected to the Data Aquisition unit via a HP-IL link. The HP-71B computer includes a quartz crystal clock and can turn on the HP3421A unit, trigger to scan a list of up to 30 inputs, instruct a mass memory to store the readings and then power down the entire system until the next time passes. The HP-9114B portable Microfloppy shall be used as mass storage media.

The active system shall be supplied by 12VDC. The peak power consumption is estimated to be less than 200 W with an average power consumption less than 25 W.

The data scanning of the required measurements will be done on a periodic basis with a time interval of the order of about 10 seconds.

By an appropriate subroutine running in the HP-71B computer the average values of the sampled measurements will be calculated and stored in the computer's RAM every 10 minutes. The 10 minute averaged values will be transfered from the computer's RAM to the floppydisc every hour.
The floppydisc will be transferred in Athens every month and from a central system a copy of a floppydisc will be produced and sent to CEC/Ispra. The analysis of data for

several useful conclusions will be done in Athens in a small central system (H-P86B)

8. SINGLE HOUSE SUPPLY

The individual house supply systems are installed with their own DC equipments. The solar generator is built up as one electrical group.

The electrical systems are very simple, therefore no electronic equipments is installed.

Only shunt regulators are foreseen. They regulate the battery voltage above the gassing voltage in order to prevent an overcharge and to reduce water losses of the electrolyte. The voltage range accuracy is +/- 2%. A temperature compensation is integrated.
Special consumer appliances will minimize the total power consumption.

A deep discharge of the battery is prevented by switching off the consumers if the voltage will fall below a certain value.

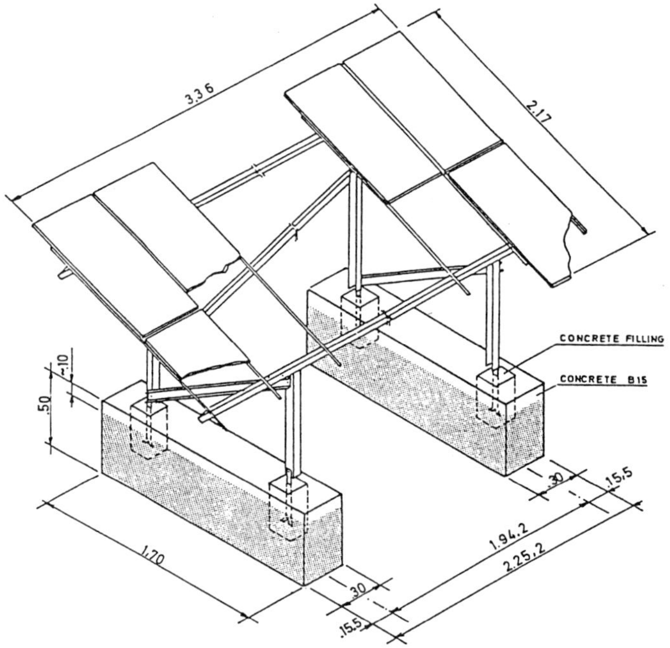

FIG. 1 ANTIKYTHIRA PV PLANT 25 KWp
PANNEL SUPPORTING STRUCTURE

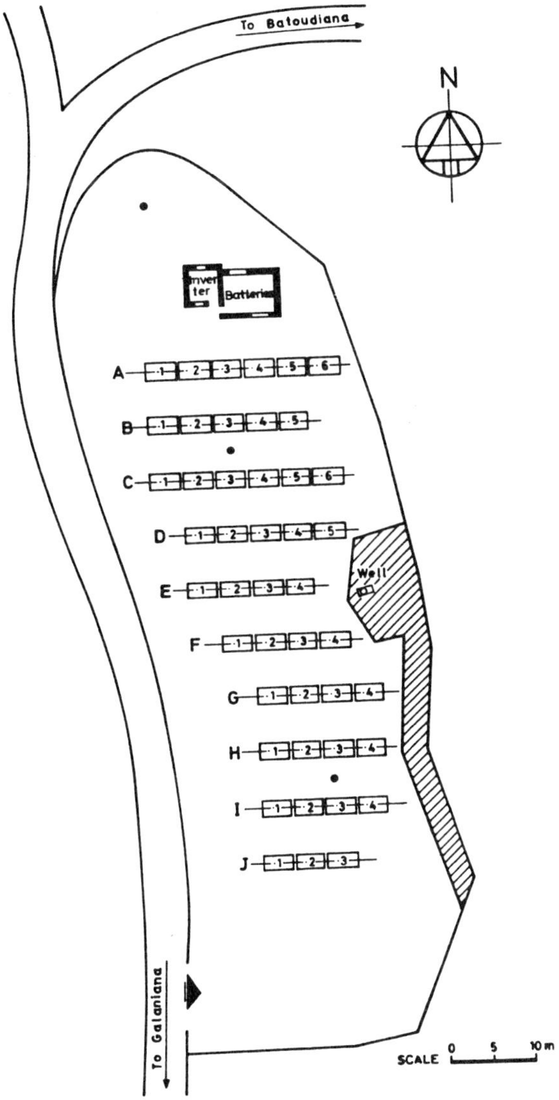

FIG. 2 ANTIKYTHIRA PV PLANT 25 KWp

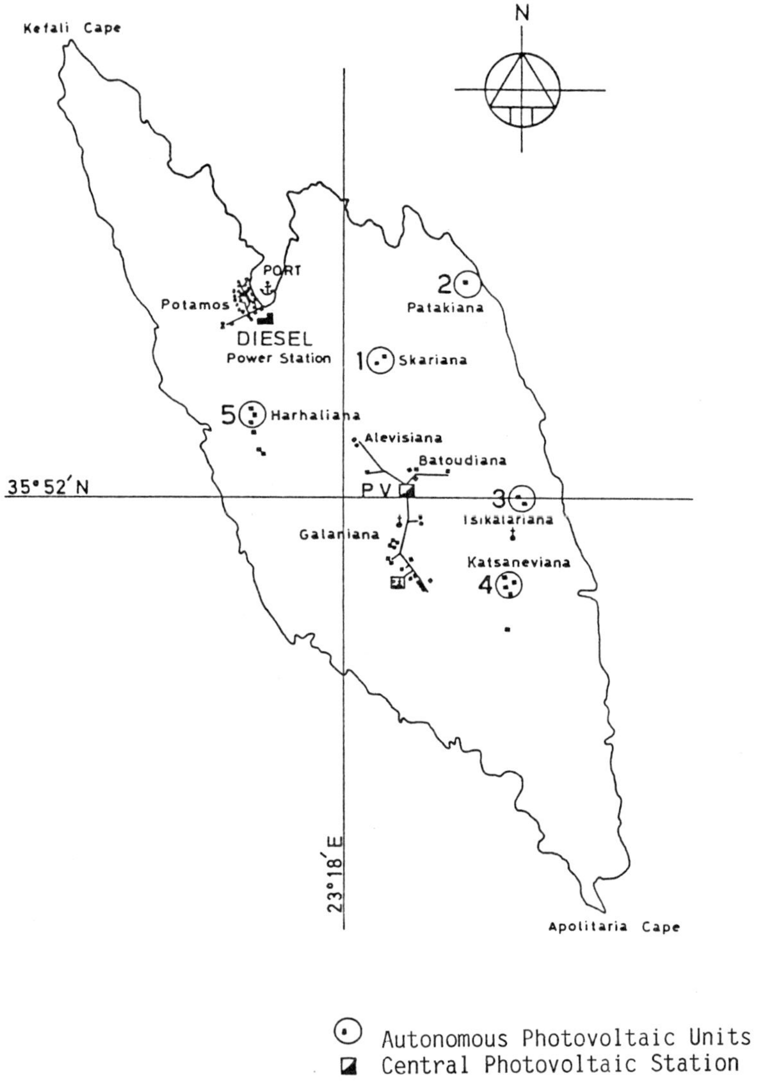

FIG. 3 ANTIKYTHIRA ISLAND

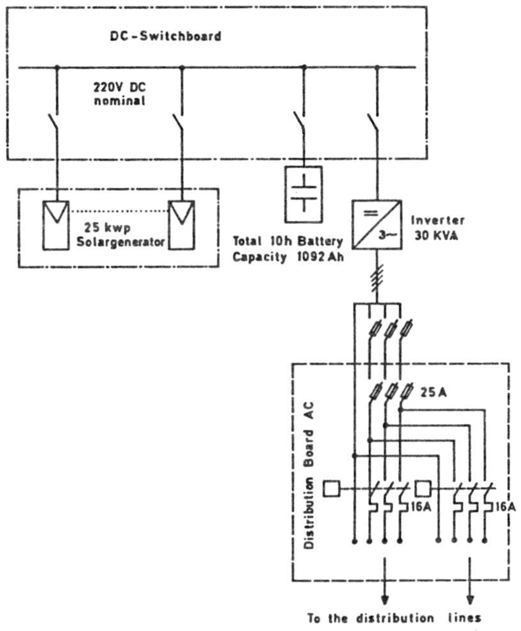

Fig. 4: Antikythira PV Power Plant 25 kWp
Bloc Diagram

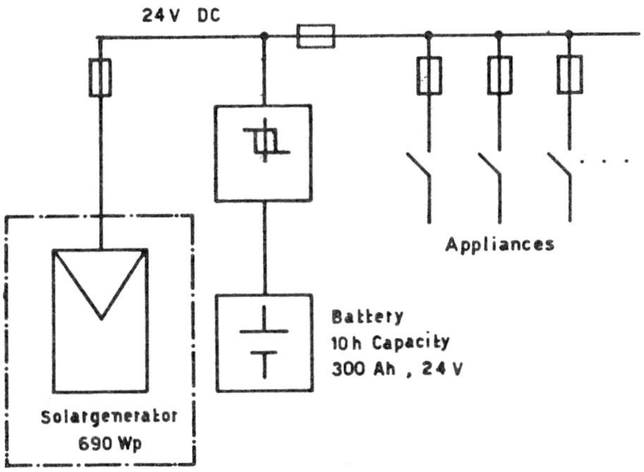

Fig. 5: Antikythira-Single House Supply
Bloc Diagram

CONTRACT NR SE/404/83

ARKI SOLAR PHOTOVOLTAIC POWER PLANT

J. CHADJIVASSILIADIS * and J. WILLE **
* Public Power Corporation
** Tractebel, Belgium

SUMMARY

The installation of a photovoltaic power plant on the Arki island aimes at supplying electricity to the community of this small and isolated Aegean island. The ultimate objective is to prove the reliability of P/V systems in autonomous operation, on which a more substantial P/V systems utilization policy could be based. A central photovoltaic plant of a capacity of 25 kWp will supply electricity to the community. High efficiency domestic appliances will be used by the consumers. All pieces of equipment of the plant have already been manufactured and tested at the factory. All of them have been shipped and are now stored in Athens. Site work is now going on, the plant will be installed close to the community. The remoteness of the site delayed the installation phase as it appeared to be difficult to find a contractor willing to execute the civil works.

1. INTRODUCTION

The project concerns the installation of a 25 kWp photovoltaic power station in the Aegean island of Arki for the electrification of its small community, (see fig. 1).

The population of the island varies from 60 inhabitants in winter season to about 100 inhabitants in summer season. Low energy cosumption domestic appliances will be used by the consumers.

2. PHOTOVOLTAIC ARRAY

The rated power of the photovoltaic array is 25 kWp. The polycrystaline silicon cells have been manufactured by Photon Technology.

Each module consists of 40 cells with a rated peak power output of 40 Wp. The whole generator consists of 40 branches with sixteen modules per branch. Each branch is connected by underground cables to the DC busbar, (see fig. 2).

The design of the supporting structures is based on standard profile sections of hot galvanized steel.

The tilt angle of the module is season adjustable, for summer time the angle is fixed at 15 ° and for the winter time at 50 ° in order to match the load requirements and to increase the power output of the PV array.

3. POWER CONDITIONING

The main objective of the power conditioning and regulation systems is to use as much as possible the available solar energy, to prevent overcharging of the batteries and to deliver the current required by the load.

The regulator consists of the regulator control unit and the regulator switching circuits, (see fig. 4). The 40 branches of the solar field are considered as current sources and are switched in or out by means of the switching circuits controlled by the regulator control unit, depending on the state of the battery and the load.

If the battery is completely charged, the regulator switches on only the branches necessary to supply the current to the inverter input in order to meet the load requirements at that time. The division of the PV array in several branches allows a very accurate control of the charging cycle of the batteries. Similarly the control unit prevents excessive battery discharge. Manual operation allowing the switching in or out of any branch is also possible. The regulator has been manufactured by CET (Liege).

The storage battery consists of 105 lead acid elements manufactured by TUDOR HELLAS. The nominal voltage of the battery is 210 V and its nominal capacity 1400 Ah.

The inverter is a 30 kVA, high efficiency three-phase self commutated inverter manufactured by AEG, operating in stand-alone mode.

The plant is also equipped with an AC/DC rectifier also supplied by CET which allows to charge the battery from an AC generating unit.

This rectifier will allow to charge the batteries form on external generating unit, e.g. installation of a wind generator has been considered in order to cope with future load increases.

4. INSTALLATION OF THE PLANT

All the equipment have been manufactured and tested at the factory. All of them have been shipped and stored in Athens ready for transportation to the site and for installation.

Due to the difficulties to find a contractor for the civil works in this remote area the installation phase has been delayed. Now the civil works are under construction and it is expected that the plant will be commissioned next September.

An extended distribution grid will be installed during summer to supply electricity to all the houses, the shops and the public offices of the community.

5. USER APPLIANCES

Low energy consumption appliances will be used by the inhabitants.

- special lamps of 18 W, 900 lm for the houses and street lighting
- refrigerators of a capacity of 120 l,70 W motor
- TV sets color TV of 50 W and 16" tube

In addition to the houses some shops and public offices will be electrified.

6. MONITORING SYSTEMS

A data aquisition system has been designed according to the guidelines of Ispra. The data aquisition and instrumentation equipment is supplied by Hewlett - Packard.

The H-P 3421 A data Aquisition unit is equipped with 30 analog and/or digital indication inputs circuits. The H-P71B Handheld computer is connected to the Data Aquisition unit via a HP-IL link. The HP-71B computer includes a quartz crystal clock and can turn on the HP3421A unit, trigger it to scan a list of up to 30 inputs, store the readings to a mass memory and to power down the entire system until the next scan. The HP-9114B portable Microfloppy shall be used as mass storage media.

The active system shall be powered by a 12VDC power supply. The peak power consumption is estimated to be less than 200 W with an average power consumption below 25 W.

The periodic scan rate is fixed at 10 seconds.

By an appropriate subroutine running in the HP-71B computer the average values of the sampled measurements will be calculated and stored in the computer's RAM every 10 minutes.

The 10 minute averaged values will be transferred from the computer's RAM to the floppydisc every hour.

The floppy disc will be monthly collected and transferred to Athens where a copy of it will be produced and sent to CEC/Ispra.

Subsequent analysis of data shall allow to derive from these measurements several useful conclusions. This analysis will be performed in Athens on a small central system (H-P86B).

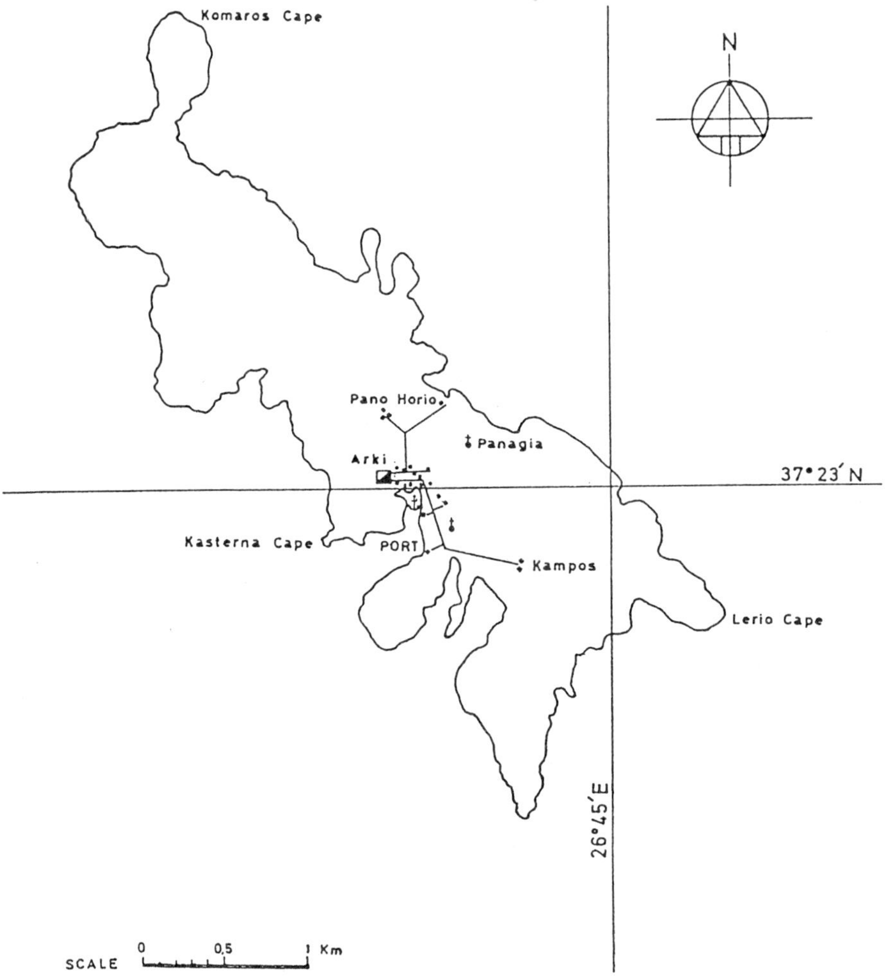

Fig. 1 ARKI PV PLANT 25 KWp
MAP OF THE ISLAND

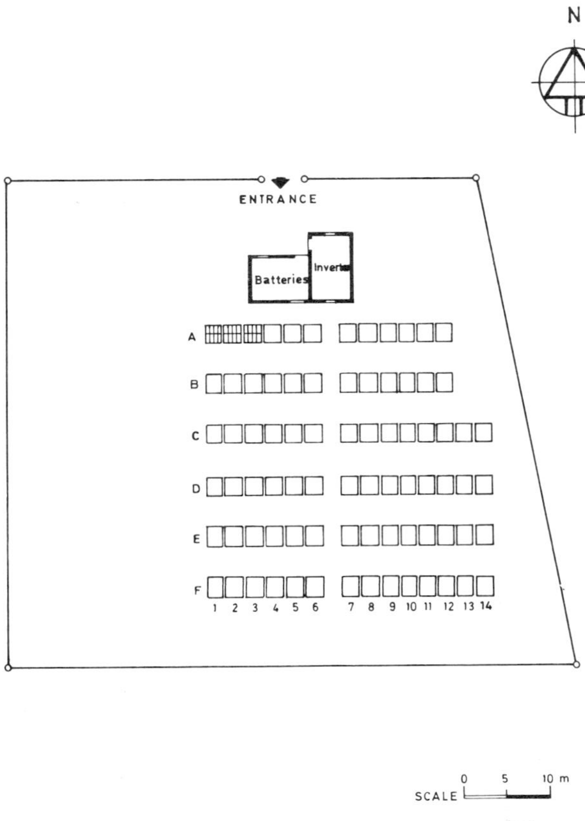

Fig. 2 ARKI PV PLANT 25 KWp
GENERAL LAYOUT

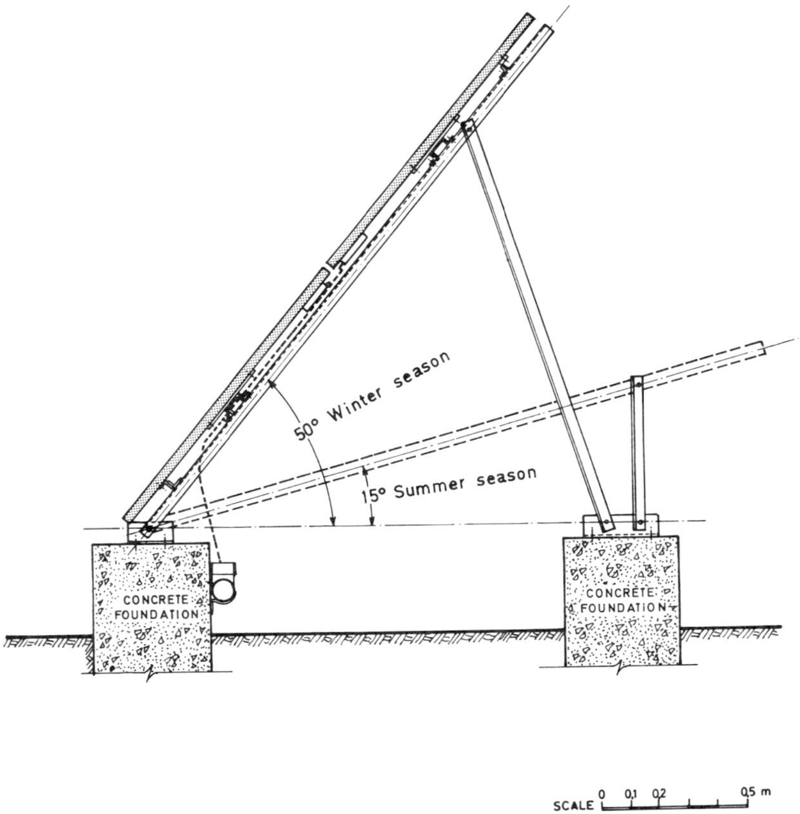

Fig. 3 ARKI PV PLANT 25 KWp
PANNEL SUPPORTING STRUCTURE

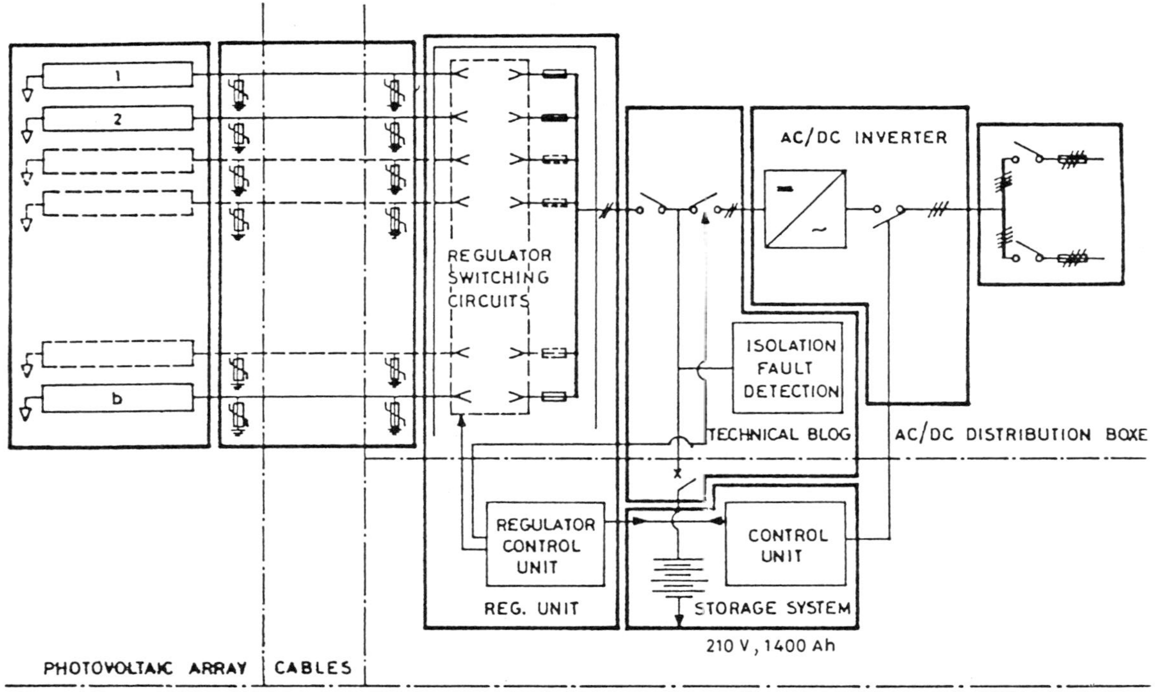

Fig. 4 ARKI PV PLANT 25 KWp
BLOCK DIAGRAM

CONTRAT NO SE-555-84-DK

SOLAR- AND WINDPOWER PLANT IN HASLE ON THE ISLAND OF BORNHOLM

H.C. RASMUSSEN+ and O.S. NISSEN++
Hasle kommune. Toftelunden 1, DK-3790 Hasle
+Consultant. Skröbelev Kirkevej 18, DK-5900 Rudkøbing
++Manager. Energy Laboratory, Niels Bohrs Alle 25, DK-5230 Odense M

Summary

The project still being in the planning and predesign stage covers a combined photovoltaic solar- and windenergy power facility for some larger municipality consumers. The plant shall be composed of

- a photovoltaic field of a nominal power of 36 kW_p
- two windmills each equipped with 99 kW generators with a total wind swept area of 600 m^2
- a battery storage system of 200 kWh
- a DC/AC inverter/power conditioning system
- a super control system.

The estimated energyproduction totals 496 MWh/y. The barrier to be removed before the actual construction takes place is tied to discussions upon the extent the project will promote Danish industry in the new field of photovoltaics.

1. INTRODUCTION

In Hasle municipality on the island of Bornholm (50.000 inhabitants) in the Baltic sea, the siting being shown in figure 1, many initiatives have been taken to establish renewable energy plants (1).

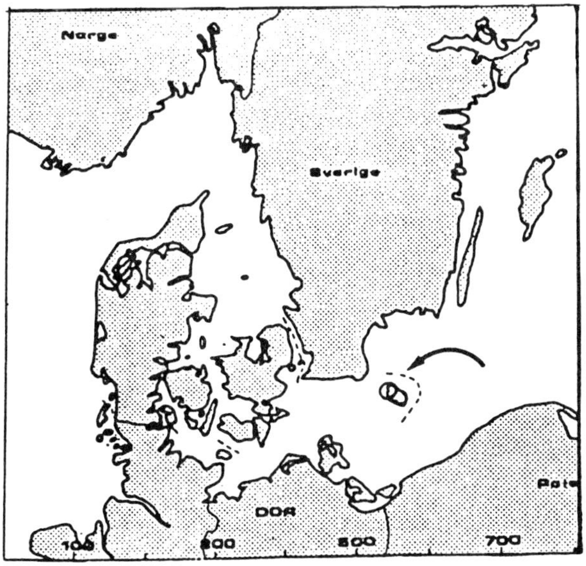

Figure 1 : Map showing Hasle on Bornholm in the Baltic Sea.

Among these is the EEC supported solar- and windpower plant. The plant construction is in the investigation and planning phase and shall be composed of

- two photovoltaic production units built with solar cell arrays in one field with a total nominal power production of 36 kW_p (AM 1.5)
- two windpower production units each equipped with 99 kW/25 kW generator capacity (300 m^2 wind swept area each, hubheight 24 m)
- a battery storage system of 200 kWh
- a DC/AC inverter/power conditioning system and a supercontrol system.

A diagram of the combined plant is shown in figure 2.

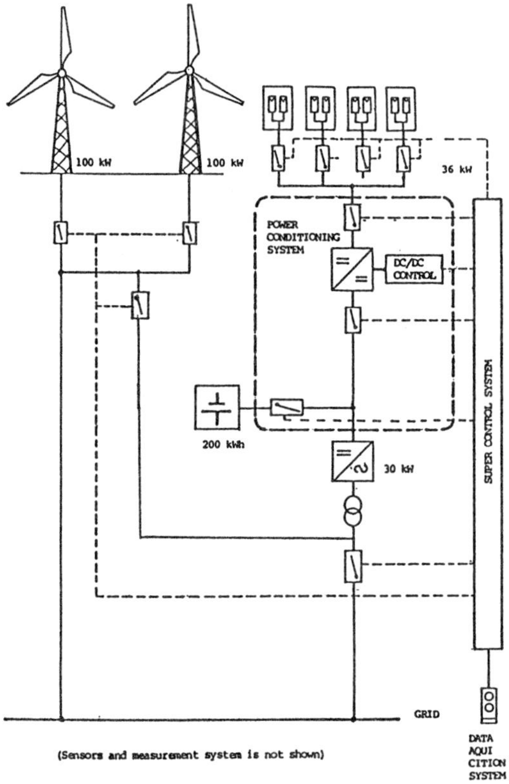

Figure 2 : Solar- and windpower plant. Block diagramme showing components and energy and information links.

The annual energy balance for the plant (norm year)	
is total	1786 GJ/yr (496 MWh/yr)
- origin wind energy	1656 GJ/yr (460 MWh/yr)
- origin solar energy	130 GJ/yr (36 MWh/yr)

Concerning the windenergy and systems equipment Danish companies are fully able to respond to the forthcoming call for tenders. Concerning the photovoltaic part first by now a Danish industrial production facility is under construction. Therefore only 25 % of the installed PV-power is foreseen to be Danish. This paper describes in more detail the functioning of the PV-part in the whole system.

2. PV-ARRAY

Figure 3 shows the appearance of the plant in the surroundings. The overall criteria for selecting the site of the plant have been :

1. The ground area for the photovoltaic units has to be plain or slightly sloping south.
2. There shall not be physical appearances in the surroundings that can shade the solar panels.
3. The whole plant shall be as close as reasonable to the nearest sub-station in the public electricity mains.

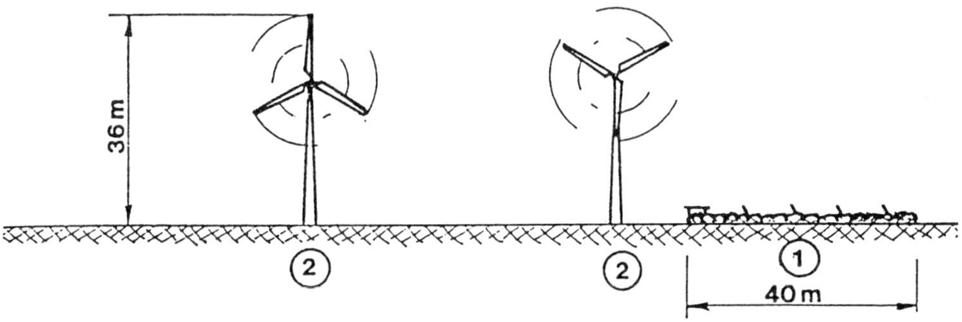

Figure 3 : Solar- and windpowerplant. Architectural appearence in the surroundings. 1 Photovoltaic part, 2 Wind energy converters.

The appearance in the surroundings are 450 m^2 of photovoltaic surface area within 2800 m^2 fenced area. The wind energy converters are situated outside the fenced area in due distance and harmonized with the nearby windpark of 9 windmills. The PV-panels are ordered in 4 individual rows. Within the area the 10-12 m^2 technical building is established. The building gives shelter to the powerconditioning system, inverter, battery bank, super control system, measuring and reporting equipment.

The photovoltaic solar energy converters (PV-SEC) consist of two production units in one field, the units being :

Unit I. Modules made of polycrystalline Si-cells of nominal 27 kW_p (AM 1.5)
Unit II. Modules made of monocrystalline Si-cells of nominal 9 kW_p (AM 1.5)

PV-SEC unit I

Three of the plant-subarrays consist of high efficiency polycrystalline silicon cells, connected in series and encapsulated and protected with high transmission glass into modules delivering 12 V output and 40 W (AM 1.5). The 672 modules are mounted in panels of 4 modules. Each panel is mounted in a rigid construction (length 1.90 m, width 1.06 m, based on Photowatt technology), given a tilt angle of 60° and at the same time ensuring a perfect behavior under the most severe meteorological situations. The arrays are ordered in switchable groups controlled by the power conditioning system to deliver a voltage of 220 V. The PV unit I of 340 m^2 array surface is to be delivered by Photowatt Int. s.a., France, through their Danish representative. The rigid structure and the connections etc. are to be manufactured locally.

PV-SEC unit II

The plant subarray consists of high efficiency mono-crystalline silicon cells prepared by a new technology developed by the Laboratory for Semiconductor Technology, The Technical University of Denmark (2,3) and manufactured by a Danish producer.

The cells will be encapsulated and mounted to give the same technical and visible appearance as the unit I cells. Thus 85 m^2 array surface will be added to give a total plant array of 425 m^2 and a total nominal power of 36 kW_p (AM 1.5).

The Danish cells will consist of 4" (80 cm^2) cells with an expected efficiency of 12.5 %. The power production will then be 1 kW_p per cell. By a new masking technique it has been possible to enhance the active area by 4 % without decreased shuntresistance. To make the production competitive in the market the cover glass is expected to be the same ironfree glass as found reasonable by Jet Propulsion Laboratory (4). As back cover 100 micron Tedlar from DuPont will be used.

The local climatic conditions of high air humidity and light corrosive properties (the plant being situated near the sea) will be evaluated. It is expected to be sufficient to provide the panels with metal rims around the edges, tightly fitted and filled with silicone rubber.

Array supporting structure

The aluminium supporting structure for the photovoltaic units has been designed by a local manufacturer. In this design a manual change of the tilt angles of the panels is possible. A sketch is shown in figure 4.

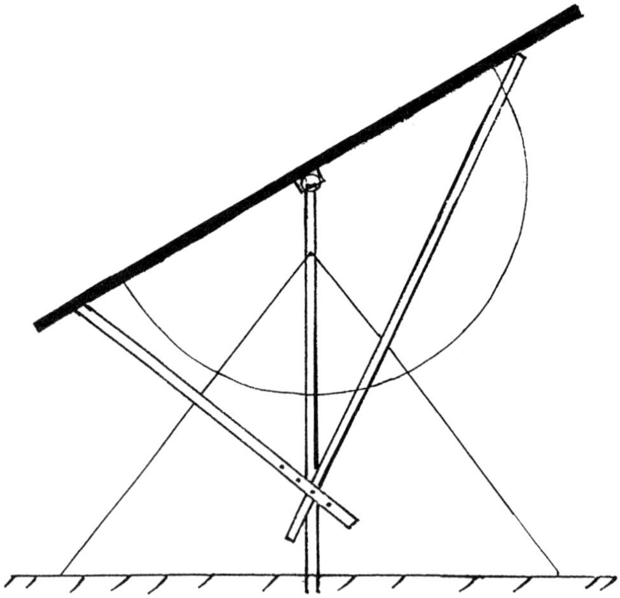

Figure 4 : Sketch of aluminium supporting structure for the photovoltaic modules.

Power conditioning

The aim of the power conditioning system (PCS) is to assure maximum power production from the panels. This is established by the PCS incorporating a DC/DC regulator unit for optimal load on the panels together with an output power corresponding to the battery bank characteristics and the inverter input requirements. In case of malfunctions the PCS switches off the arrays necessary.

The battery capacity assures that on clear days excess production is stored for later night use. Since the excess power from the PV-SEC is less than 10 kW and only appears in a few hours during summer days, only a small battery bank of 200 kWh is necessary. The charging of the battery will be done exactly on the charging curve alternatively by switching over the corresponding numbers of PV-SEC groups, thus assuring a prolonged lifespan of the battery. The nominal discharge power being selected to 30 kW at a voltage of 187-264 V. Accordingly, deep discharge is prevented by the PCS.

DC/AC inverter

The DC/AC inverter allows to generate alternating voltage from a DC voltage. The inverter input is delivered either by the PV-SEC or by the battery controlled by the PCS. The inverter will be a static 30 kVA three phase one with fully controlled thyristor bridges. It accepts a DC input voltage of 220 V +- 25 % with an output fully meeting the utility standard of 3 x 220 V/380 V, 4 wire, 50 Hz +- 0.5 %, 0/380 V +- 2 %, 0/220 V +- 1 %.

Super control system (SCS) and data acquisition system

Data from all mainparts of the system are fed to the SCS and by a small computer analysed. Control is made to assure overall maximum power production and matching to the load. The computer also stores all main data for later data analysis.

4. CONSUMER INVESTIGATIONS

The power demand of the whole island of Bornholm has earlier been reported. For two typical days the curves are shown in figure 5 (5).

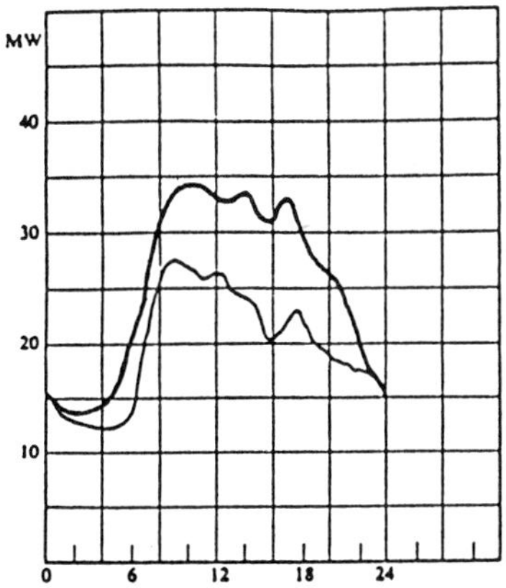

Figure 5 : Power demand on Bornholm on a typical summer day (16 June 1982) -lower curve- and winter day (15 Dec. 1982) -upper curve- respectively.

The demand of the main Hasle municipality consumers during one year has been investigated to identify suitable loads for the solar and windpower plant. Of the loads given in table I further investigations shall be performed as to power variations during the year and during the day.

Electricity demand (1984)	MWh/yr	GJ/yr
Hasle public school	51.8	186.5
Hasle public school, sports facilities	108.3	389.9
Old Peoples home, Toftelunden	263.5	948.6
Old Peoples residence, Gammeltoft	600.0	2160.0
Hasle Harbour	75.0	270.0
Municipality water works	175.9	633.2

Table I : Larger municipality electricity loads in the town area of Hasle.

5. MONITORING SCHEME

The measuring programme have the objectives

- to collect the information necessary to adjust the plant and the SCS
- to make calculations of efficiencies of all energy conversion processes possible
- to correct and improve the model used in designing the plant
- to improve PV-technology in general through delivering performance data to the ISPRA-establishment every two months
- to document the load matching improvements obtained by combining wind-, solar electricity and energy storage compared to individual-source-plants without storage.

All measurements will be performed in accordance with the guidelines set out by Blässer and Krebs (6). The measurements include (one minute interval and averaged every hour if not otherwise stated)

- global insolation
- insolation on the array-plane
- wind speed at 10 m height and at WEG-hub-height (10 min. average every hour)
- wind direction at 10 m height
- max. ambient temperature
- module temperature
- array output energy - unit I
- array output energy - unit II
- converter output energy (DC)
- energy to battery from PV-part
- energy to battery from WEG
- energy from battery
- battery voltage (every 6 hours)
- energy from WEG (1)
- energy from WEG (2)
- energy to the grid
- energy to public consumers load (every hour).

Included will be the investigation of an IBM-PC compatible data acquisition system for the measuring campaign.

6. DISCUSSION

This future construction of a combined solar- and windpower plant is not regarded as a prototype for the near future power supply of the larger Danish islands connected to the mains of the large power stations. Though it gives an increased local supply security, this cannot balance the necessary investments. What shall be achieved however for Danish and EEC industries is experiences with a power source of tomorrow allready competitive in many parts of the world.

The benefits to be obtained from this particular plant situated on a (remote) island are to give experiences from quite northern latitudes meaning a low insolation in wintertime on solar energy combined with wind power and a battery bank to become a more stable power source.

REFERENCES

(1) RASMUSSEN, H.C. : System analysis of a solar- and windpowerplant for Nordic remote locations. Energy for rural and island communities IV. Proc. Fourth Int. Conf. Inverness. Scotland 16-19 Sept. 1985. Eds. J. TWIDELL, I. HOUNAM and C. LEWIS. Pergamon 1986 pp. 45-52.

(2) LEISTIKO, O., Y. SAFIR and W.G. PROCTOR : $P^{+}NN^{+}$ and PIN low cost silicon solar cell technology. 17th EC Photovoltaic Solar Energy Conference 1986.

(3) SAFIR, Y., O. LEISTIKO and W.G. PROCTOR : Low cost solar cell technology. Photovoltaic Solar Energy Conference II, 1986.

(4) JPL-81-102. Photovoltaic module encapsulation. Design and materials selection. Vol I. 1981.

(5) ØSTKRAFT : Beretning 1983. Rønne 1984.

(6) BLÄSSER, G. and K. KREBS : Guidelines for photovoltaic system monitoring. ESTI, EC, Ispra 1985.

SESSION III - OTHER PROJECTS

SESSION 3 : OTHER PROJECTS

3.1 OVERVIEW OF SESSION 3

3.1.1 Lighthouses

There are currently 4 major lighthouse projects in the demonstration programme. Presentations were made on the following 1983 and 1984 projects.

(SE/624/84) Lithari Lighthouse

Mr Kadella from Siemens Solar (formerly Interatom) reported on progress with this project to electrify the Lithari lighthouse on the Greek island of Skyros. A 4.7 kWp PV generator will power a static beacon with 3 beams. The energy consumption of the lamp for an operating period of 14 hours is 8.1 kWh, and power for 10 days can be supplied by a set of batteries with a capacity of 1600 Ah. An analytical monitoring system will be powered by an additional PV generator of 650 Wp.

(SE/648/84) Methoni Lighthouse

Mr Kadella described this project to electrify Methoni lighthouse on the Greek island of Sapientza. The design of this system is now complete and installation should be finished at the end of the summer of 1987. Access to the island is difficult and is only possible by boat in good weather.

A 1.3 kWp generator will power a rotary lamp, which has a range of 22 nautical miles. Additional power will be supplied by a small wind generator rated at 1 kW. The lamp has an energy consumption of 1.9 kWh for a 14 hour operating period, and 12 days of back up will be supplied by batteries with a capacity of 1000 Ah.

A small global monitoring system will be installed, which will be powered by a 30 Wp PV module.

(SE/674/84) Palmaiola Island PV Lighthouse

Ms Orsoni from Pragma reported that this PV powered lighthouse was now in operation. Installation of the 4.7 kWp main array and 0.4 kWp emergency power array was completed in September 1986. The main array charges a 120V, 600 Ah battery, and the emergency array a 12V 100 Ah battery.

The lighthouse, which is owned by the Italian navy was previously gas powered. The PV elecrification of the lighthouse will increase the operating range of the lighthouse lamp from 14 to 21 nautical miles and lower maintenance and man power costs.

Access to the site is a problem, and equipment had to be flown in by helicopter due to a lack of suitable landing places and a steep lane from the beach to the lighthouse. Analytical monitoring of the system began in October 1986 and the first diskette of data will be collected in May.

Bird spikes have been mounted along the edges of each of the PV modules in order to prevent seagulls from perching on the modules.

3.1.2 Other Applications for PV Systems

Other applications for PV generators in the demonstration programme, are varied and include telecommunication systems, warning systems, monitoring systems and electrification of remote sites. The following 1983 and 1984 projects were discussed.

(SE/704/83) PV Installation of the RF Station Antikythira

Mr Kadella from Siemens Solar reported that this PV plant had been working well since it was commissioned in October 1985. A 20 kWp PV generator provides power for a telephone, radio and television relay station on the Greek island of Antikythira. Two diesel generators are available for back up but normally only run for about two hours a month. The starter batteries of the diesel generators are kept charged by 3 small PV modules.

A Hewlett Packard data acquisition system is used for analytical monitoring of the system. A full set of data was not available for some months, but this was thought to be due to a failure of the data acquisition system rather than a failure of the PV generator. It was suggested that a simple electro.mechanical counter be installed so that if the data acquisition should fail, then it would still be possible to determine whether or not the PV generator was still working.

The project has been undertaken in collaboration with OTE, the Greek telecommunications organisation, and at the start of the project training was given to several staff from OTE. Unfortunately these staff are no longer available to work on the project, so there have been delays before problems with the PV generator or data acquisition system can be solved.

(SE/123/83) PV Generator to Supply an Early Warning System for Forestal Fire

Mr Fonzi from Italenergie described the two PV powered warning systems for detecting forest fires and for predicting frost which have been installed in the Sulmona valley and by the Pescara river. The forest fire warning system, consists of an optical unit to check smoke emissions, an infra red sensor and a steering unit to rotate and tilt the sensors. The data is transmitted back to a central computer where it is analysed. The whole system is powered by a 400 Wp PV array with a 300 Ah battery.

The frost prediction system measures temperature and humidity and transmits the data to a central computer. The occurance of frost can be predicted from the temperature and humidity at sunset with 95% certainty.

(SE/560/83) PV Power Supply for Groundwater Level Measurement

Mr Lowalt from AEG described this project to install PV systems to measure the groundwater in 3 wells near Hamburg. Two systems were commissioned in October 1986 and the third in March 1987. The installation of the third system was delayed due to objections over the choice of the site. The final site was 30 km away from the data collection centre, and a stronger radio transmitter had to be built for the system, thus further delaying installation.

Each system is powered by a PV generator of approximately 300 Wp. Data on the water level in the well is recorded continuously and transmitted to the data collection centre as required. Global monitoring of the systems has begun and data should be available soon.

(SE/267/84) PV Power Supply to the Bird and Weather Station on Sharhörn Island

Sharhörn Island is a bird sanctuary which is inhabited from March to November by one to three scientists. Mr Löwalt reported that a 4.6 kWp PV generator had been installed to provide electrical power for scientific activities. Commissioning of the PV generator began in April 1987 and end use appliances should be installed by the middle of May. The PV generator has been designed to meet all electrical needs from March to September, as this is a critical period in the nesting and hatching season, of the birds. This project may supply useful information on the behaviour of birds around PV arrays.

CONTRACT NO. SE 624 84 HE

AUTONOMOUS PV/BATTERY POWER SUPPLY FOR AN UNMANNED LIGHTHOUSE (LITHARI)

H. KADELLA *) and D.TSANAKAS **)
*) Siemens Solar GmbH, Bergisch Gladbach/FRG
**) Democritos University of Thrace, Xanthi/Greece

Summary

The existing system of the Lithari lighthouse operating with oil and the permanent manual work of 6 persons shall be replaced by an autonomous and unmanned system powered by a photovoltaic generator of 4680 Wp. The battery of a total capacity of 1600 Ah at 48 V can supply the whole beacon system for 10 days without the necessity of recharging. A large scale measuring station according to the ISPRA-guidelines for supervision of PV-systems greater than 3 kWp is foreseen for the 2 years monitoring campaign. Moreover, a telereporting system (digisponder) is integrated to transmit the operation data.

1 INTRODUCTION

The Lithari lighthouse, on Skyros Island, was built in 1850 and operates with lighting oil. Access is quite difficult while a permanent personnel of 6 persons is maintained with a minimum of 3 people in continuous attendance. The existing system comprising rotation mechanism, optical lenses, oil lamp, centrifugal regulator, shall be replaced by a new long-range electric flashing beacon system powered by a photovoltaic generator system (see Fig. I).

An adapted telereporting system transmits significant data and alarms of the facility to a central control station.

The project shall be executed in the existing lighthouse "LITHARI" on Skyros Island in Greece with the geographical data of 38° 46' latitude and 24° 40' longitude.

The intention of the project is to demonstrate the possibility of the energetic-autonomous and reliable operation of the lighthouse system powered by solar generators under the environmental conditions of the Greek coasts.

In case of success of the project, some important advantages in comparison to old conventional lighthouse systems will be available:

- Reduction of the running costs of the existing lighthouse
- Saving fossil fuel costs
- No operation Staff
- Reduction of maintenance works

2 PHOTOVOLTAIC ARRAY

Three static beacons of type PR 500 DC from Automatic Power will be installed and will be autonomously supplied with electricity by an array of solar generators of 4680 Wp.

The energy demand of one beacon is specified as 56.21 Ah at 48 V with a peak current at switch-on-moment of 300 A (nominal current 24 A).

The total energy demand of the complete system is 168.63 Ah at 48 V for a 14 h operation with a grpfl (3)-characteristic. That means that the PV-generator has to supply minimum 8.095 Wh per day to the beacon system.

To deliver this energy with a 15 % engineering supplement even in the month with the least solar radiation (December) of

$$107 \frac{kWh}{m^2 month}$$

without taking additional energy from the battery, a solar generator of 4,300 Wp has to be installed. The hardware configuration which was chosen with the aspect of certain redundancy comprises 36 solar modules instead of 33.1 modules. So an additional supplement of 8.8 % is achieved.

The energy which can be delivered by the battery is essentially determined by the peak current of the complete beacon system and the lower voltage limit of 1.75 V per battery cell. The chosen battery capacity of 1.600 Ah at 48 V can supply the beacon system for 7 - 8 days with energy without being recharged, if the peak current of the complete beacon system is limited to approximately 350 A. This can be done by a delayed (approx. 0.1 sec. from each other) switching-on of the 3 beacons. If the peak current was 900 A, the battery capacity would only be sufficient for 3.5 to 4 days.

Data of PV-generator:

Daily power consumption	8,100 Wh/d
Minimum insolation:	107.9 kWh/m^2/month
Amb. panel temp. during operation:	25 °C
Nominal solar generator voltage:	48 volt
System efficiency:	77 %
Engineering security supplement:	15 %
Minimum power of solar generator:	4,300 Wp
Equiv. quantity of modules SM 144:	33.08

Array comprises 36 modules SM 144-18; 9 parallel strings of 4 modules in 1 line.

3 POWER CONDITIONING

DC-Chopper

Type:	GSH48/6800
Manufactured by:	Interatom

Characteristic features:

- Microprocessor controlled power conditioning unit
- Modular constructed with one central control section and 1 to 4 power sections each of 1700 W maximal
- Available power range up to 6800 W (1 - 4 power sections)
- Maximum power point (MPP) regulation for maximum power extraction from the solar modules for each temperature
- High efficiency ($\geq 93\%$)
- Following data can be read out from central control board via IEEE 488 or RS 232c interface:

  Input voltage and current

Output voltage and current
Temperature of the cooling ribs and battery

- Reduction of battery overcharge limit according to battery temperature
- Sensor input for battery voltage

Technical data:

Rated power per device	6800 W
Rated power per unit	1700 W

Battery Storage for PR 500 system

Type:	Varta Bloc
Manuafactured by:	Varta AG

- capacity 1600 Ah at 48 V
- Expected service life is approximately 10 - 15 years, depending on conditions (especially temperature).
- The self-discharge rate is less than 3 % C_{10} per month at 25 °C.

4 MONITORING SYSTEM

The guidelines for supervision of PV-systems published by the Commission of European Community (CEC) require for PV-plants greater than 3 kWp an extensive data logging with specified accuracy and data storage on special media. For smaller systems only a hand-written report of daily averaged values is required.

The solar generator for the PR 500 system on the island Skyros comprises 4.68 kWp. In this case an extensive data acquisition is required. It can be done by a system which consists mainly of components which are normally designed for the use in industrial applications under rough environmental conditions. Besides the data are recorded on a tape recorder just in the form required by CEC.

The PV-Generator for Data Logging System comprises 5 parallel modules of type SM 144-36.

The daily power consumption of the data logging system is 1200 Wh/d.

A battery is foreseen for a 10 days sunless period with a capacity of 450 Ah at 24 V nominal battery voltage (see Fig. II).

5 INSTALLATION AND COMMISSIONING

The installation of the system will be during summer time, the commissioning is foreseen for middle of October.

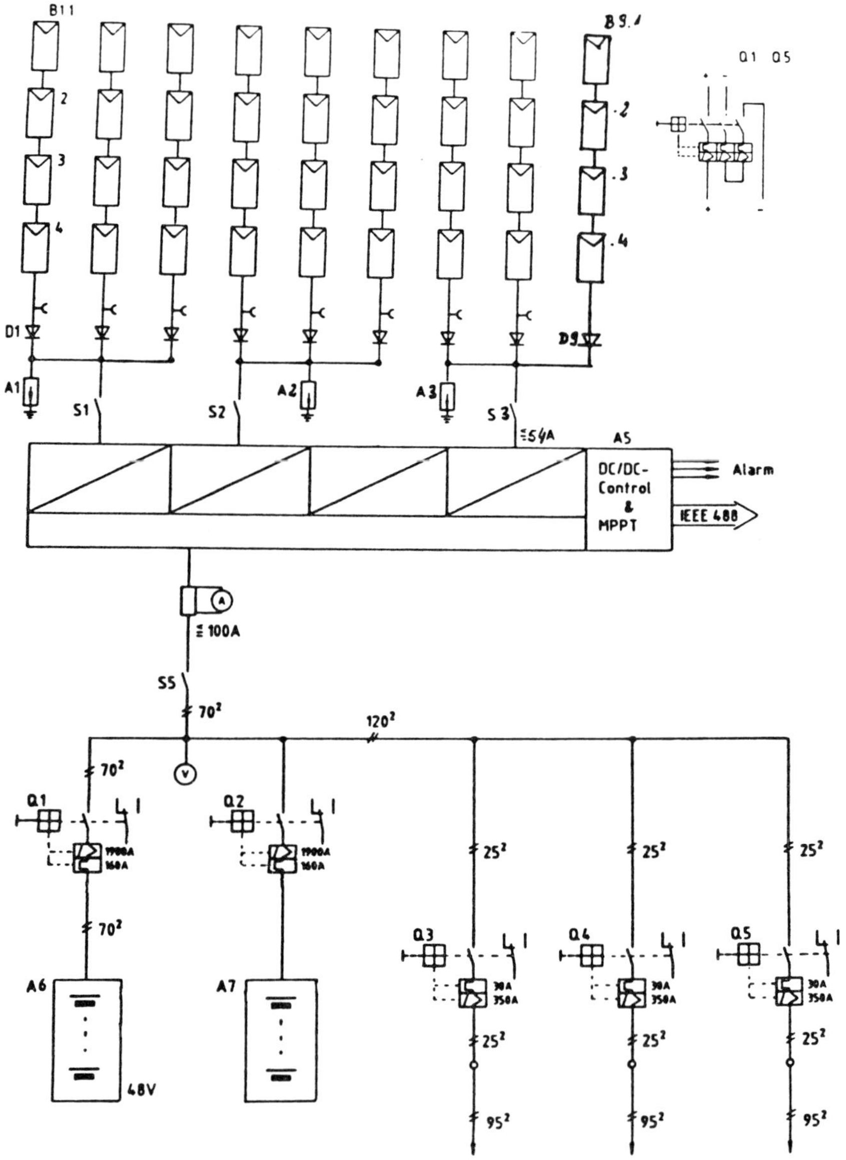

Fig. I: Power Supply for Static Beacon System PR 500-DC (Lithari)

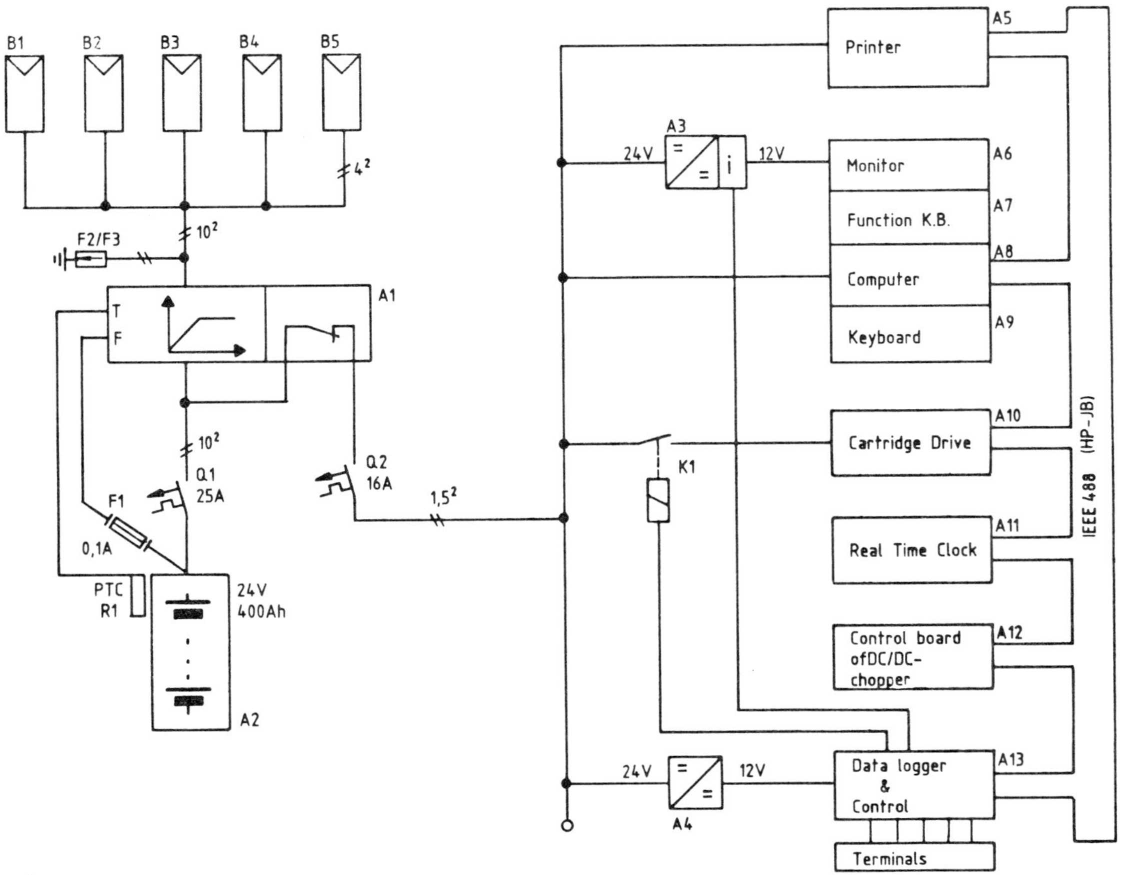

Fig. II: Data Logging System (Lithari)

CONTRACT NO. SE 648 84 HE

AUTONOMOUS PHOTOVOLTAIC AND WIND POWER SUPPLY FOR AN UNMANNED LIGHTHOUSE (SAPIENTZA)

H. KADELLA *) and J. CHADJIVASSILIADIS **)
*) Siemens Solar GmbH, Bergisch Gladbach/FRG
**) Public Power Corporation, Athens/Greece

Summary

A new long-range flashing system, rotary type, shall replace the existing system which consists of beacon, oil-lamp, supply system and rotation mechanism. The beacon shall be electrical, autonomously powered by a combined solar photovoltaic and wind generator system. Both systems feed the energy in a battery storage. The rated power of the solar generator is 1300 Wp. The battery of a total capacity of 1000 Ah at 24 V can supply the whole beacon system for 12 days without recharging. An automatic telereporting system is integrated to transmit data and alarms of the facility to a central control station. Moreover, an autonomous photovoltaic powered data monitoring system according to the JRC-Ispra-"Guidelines for Photovoltaic Systems Monitoring" shall be installed for the 2 years monitoring campaign.

1 INTRODUCTION

Along the Greek coasts operate today about 900 lighthouses and beacons. It is historically verified that their installation started in about 1831 and has gradually increased to the present number with projections to be extended up to twice as many according to international navigation rules.

The first beacons were of a rotating type with a glass-lense system and a glowing type lamp using lighting oil as fuel, reaching a range of up to 22 nm.

The Sapientza lighthouse was built in 1850 and uses an optical element of 1.85 m in diameter and operates with lighting oil.

The lighthouse is placed at a height of 96 m above sea level, on an inaccessible location. Approach is quite difficult while a permanent personnel is responsible for the lighting and extinguishing of the beacons, the immediate intervention in case of irregular operation as well as the winding of the lighthouse's rotation mechanism. The main part of this mechanism is composed of a system of gears, which in turn, rotates the ring supporting the optical lenses via counterbalances and a centrifugal regulator. The interval between two successive windings is 1.5 hours.

The aim of this project is to install a new long-range electric flashing system, rotary type, in the existing lighthouse.

The beacon shall be autonomous, electrically powered by a combined photovoltaic and wind generator system.(see Fig. I).

The solar modules will operate via battery charge controller on a 24 V battery storage of 1000 Ah.

The beacon DSL 3 with a grpfl (3) characteristic in a 30 seconds period and a luminous range of 23.2 nm at transmission factor 0.74 needs an energy of 1890 Wh per day.

Engineering works have started with the acquisition of environmental and site data that are needed for the final design of the complete system.

2 PHOTOVOLTAIC ARRAY

The ten solar modules are divided in 2 subarrays and the battery consists of two parallel groups. So a certain redundancy can be achieved with respect to particular system failure.

The solar generator sizing is done with the insolation values of Athens which are available from a data volume of University Wisconsin. The values can be taken as reference for the Sapientza island. Generally the insolation values on this island are better than in Athens.

The solar generator is designed to supply the system also during the month with the least solar radiation (December) of

$$107 \frac{kWh}{m2\ month} \ in\ 60^{\circ}\ tilt\ angle$$

the energy demand of the complete marine rotary beacon for a 14 hour operation.

Daily power consumption:	2.124 Wh/d
Minimum insolation:	107.9 kWh/m 2. month
Amb. panel temp. during operation:	20 deg. C
Nominal solar generator voltage:	24 volt
System efficiency:	67 %
Engineering security factor:	1.1
Minimum power of solargenerator:	1.130 Wp
Equiv. quantity of modules SM 144:	8.11

Array comprises 10 modules SM 114-18; 5 parallel strings of 2 modules in line with a total rated power of 1300 Wp.

3 POWER CONDITIONING

Mode of operation:

The regulator is a battery charging control unit for photovoltaic power supplies. The control unit automatically limits the battery charging voltage and regulates the charging maintenance current. When the battery voltage decreases again, the charging operation is automatically started, i. e. taking the irradiance into consideration, the battery is always maintained on a very high charging level.

Diodes prevent battery discharge via the solar generator during sunless periods.

The regulator monitors the battery voltage. Separate connection terminals are available for sensor lines to monitor the battery voltage.

The maximum regulator output voltage varies inversely with temperature to match battery voltage requirements, which also vary with temperature. For this purpose, the remote temperature sensor which belongs to the offer, has to be connected to the regulator (approx. -6 mV/K for each battery cell).

The regulator recognizes the following pre-fault and fault conditions:

a) Pre-fault (Battery voltage out of range)
   - Battery voltage greater than the overcharged control voltage threshold
   - Battery voltage less than the discharge control voltage threshold

b) Fault (Load off)
   - Battery voltage less than the overdischarge control voltage

- Load current greater than the allowed maximum load current

If one of the pre-faults or faults is recognized, it is signalized with the light-emitting diode and a floating signal contact (max. 220 V, 10 A AC or 24 V, 8 A DC) is opened so that the fault signal can be externally processed.There is one contact for pre-fault and one for fault.

The individual pre-fault/fault type is indicated via the corresponding light-emitting diode:

- "Battery voltage Overcharging"
- "Battery voltage Discharged"
  "Load off Overdischarged"
  "Load off Fuse"

Batteries:

24 Battery cells Varta Block Vb 2410 are arranged on 2 parallel lines of 12 battery cells in each line. The expected service life is approximately 10 - 15 years, depending on operating conditions. The self-discharge rate is less than 3 % C_{10} per month at 25 °C. Capacity: 1000 Ah.

Battery Charge Controller

Type: LR45-24MA 1

Manufactured by: Interatom

Characteristic features:

- Protection of battery against deep discharge and overcharge by automatic load disconnection and current limitation respectively.
- Solid state shunt switching regulator using SIPMOS power transistors which force the connected modules in short-circuit operation.
- 100 Hz operating frequency with defined current gradients
- Shottky diode as blocking diode
- Load disconnection with optical signal, when the battery voltage is below the deep discharge threshold (adjustable) and automatic reconnection with hysteresis
- Temperature depending reduction of maximum charge voltage by PTC-sensor input of battery temperature

Switching Regulator:

Type: PSR 128-7 Option: CO3/M/L/P

Manufactured by: Melcher AG

Characteristic features:

- Voltage input 15 ... 80 V
- Output current (max.): 8 A
- High efficiency = 89 %
- Parallel operating
- Indication signal for failure (connectable to telereporting system
- Power Fat-technology
- Max. temperature TA_{max} = 71 °C

4 MONITORING SCHEME

A small meteorological station will be installed at site. A special feature is the compact design and the robustness of the used measuring devices (Fig. II).

This station can be installed at nearly every site without special protection requirements.

Data acquisition and logging is done automatically with a minimum of manual operation (exchange of micro cassette and printer paper, correction of internal clock).

Modifications and extensions are easily possible resulting in changing some parameters in the acquisiton program and installing additional sensors.

A telereporting system (Digisponder) will collect the information relating to the operation of the lighthouse and will transmit this information to a central station. This may operate without personnel and will monitor the operating condition and may interfere in case of an emergency. (It will automatically call by telephone preselected telephone numbers and will announce the emergency occurred or may when asked by telephone give a report concerning the existing situation).

5 INSTALLATION AND COMMISSIONING

The installation of the system will be during the summer time, the commissioning is foreseen for middle of October.

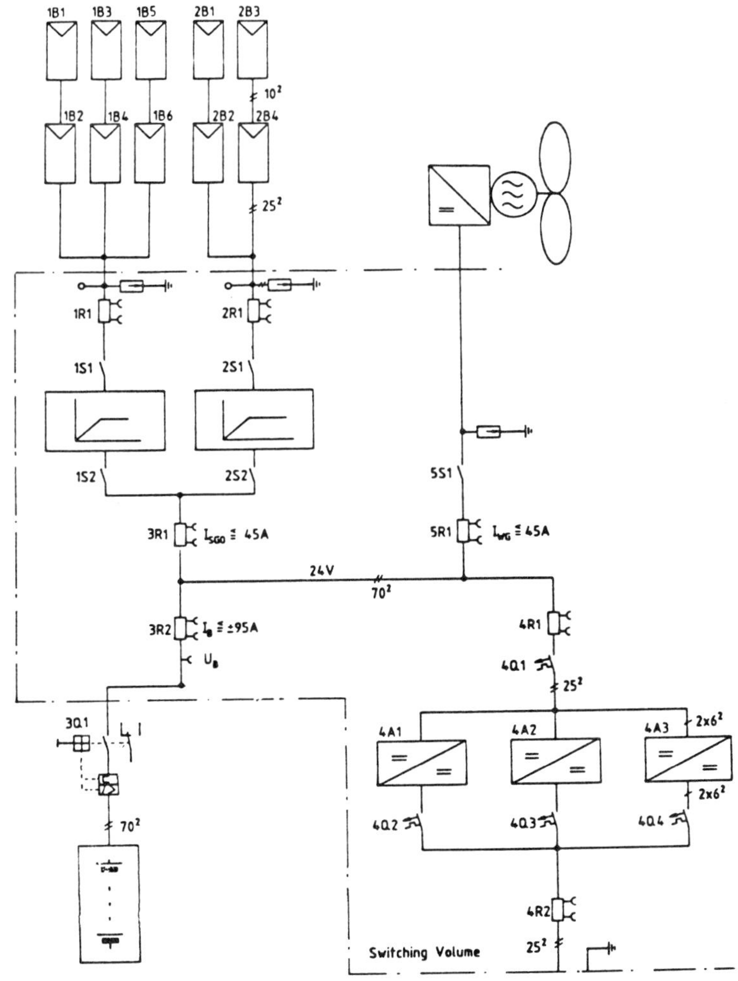

Fig. I: Power supply for Marine Rotary System DSL 3 (Sapientza)

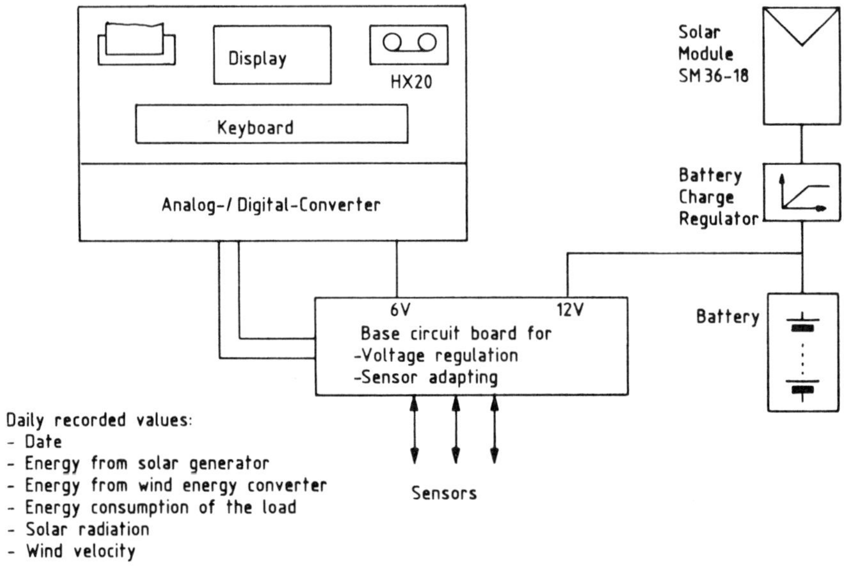

Fig. II: Small Meteorological Station (Sapientza)

CONTRACT NR SE - 674 - 84

PALMAIOLA ISLAND PHOTOVOLTAIC LIGHTHOUSE

F. Basevi, E. Bricca, M. Orsoni
PRAGMA S.p.A.
Via A D'Andrea, 6 - 00048 Nettuno - I

SUMMARY
Photovoltaic is a form of energy particulary suited when unfavourable environment makes fuel supply expensive and dangerous; gas fueled lighthouse are a typical example of abovesaid difficulties.
Palmaiola island lighthouse, placed on a small island in the Tyrrhenian sea, has been converted from gas to photovoltaic thus increasing considerably the efficiency and the operating range of the lighthouse while lowering its operating and maintenance costs.

1. INTRODUCTION

Project is aimed to supply power to the unattended lighthouse of Palmaiola island.

Palmaiola is a small island placed in the middle of Piombino channel approximately half way between Piombino and the island of Elba.

The lighthouse is property of the Italian Navy, that agreed to "convert to photovoltaic" an important navigation signalling unit with the goal of increasing the operating range of the lighthouse, from 14 to 21 nautical miles, and of lowering maintenance and operation expenses, quite heavy in lighthouses supplied by gas.

The project has been carried out from May to October 1986, in summer season, with the help of boats of the Italian Navy and a contractor helicopter. The whole system is composed by two separate photovoltaic arrays, a main array feeding the light's lamp, and a second acting as an emergency power supply.

The main photovoltaic array hangs from the south-east wall of the lighthouse, in correspondence to the windows of the upper storey. The emergency photovoltaic array is mounted on the terrace adjacent to the light tower.

In order to keep seagulls from perching on the array, lengths of high - strength steel wire (bird spikes) were mounted along the perimeter of the frames of each system's panels.

Both systems were activated on September 29th, 1986 and the Italian Navy, future owner of the system, was informed of this at the same time. A pictural view of the plant is shown in fig. I.

2. PHOTOVOLTAIC ARRAY(S)

The main photovoltaic array, feeding the light's lamp, comprises 70 Pragma 2x36 type modules (the carachteristics of which are listed below) connected in series of ten, in order to obtain a generator with a rated output of 4.5 amperes at 120 V d.c. (4,690 Wp exactly).

The emergency photovoltaic array (402Wp) consists of six Pragma 2x36 type modules, parallel connected to feed a 100Ah, 12 V battery located in the light turret, and is activated by a signal from the main control system.

Module Pragma DL 36x2 (monocristalline 4")
Electrical Data (Typical):

Open circuit voltage, Voc	21 V
Shorrrt circuit current, Isc	4.5 V
Max power output, Pm	67 W
Voltage at Pm, Vm	16 V
Average current density of cells, Isc	29 mA/cm²
NOCT	39 deg.C.

The the module characteristics dependance on temperature is reported here below (insolation = 100 mW/cm2):

	25°C	0°C	45°C	60°C
Voc. Volts	21	22.5	19.6	18.6
Isc. A	4.5	4.5	4.5	4.5
Isc, ma/cm²	29	29	29	29
Vm, V	16	17.9	14.9	14
Im, A	4	4	4	4
Pm, W	67	71.6	59.6	56

3. POWER CONDITIONING

Power conditioning and control is made through an electrical switch-board, which bears:

- terminal board
- voltage regulator
- flasher for the main light (signal)
- data acquisition unit.

Voltage regulation is achieved by interrupting the current in five of the seven string of the main array. The two remaining strings are wired directly to the battery through a clamping diode, as the USA Coast Guard standard demands for purposes of reability.

The five strings that can be disconnected are regulated as follows:

- one regulator controls three strings, and is calibrated for a cut-in threshold valueof 132 V,
- a second regulator controls the remaining two array strings and is calibrated at 138 V.

Thus, a stepped regulation is obtained, especially suitable for battery proper operation.

The two unregulated strings make up for the battery self-discharge and for the power drain of the data acquisition system.

Furthermore, they supply a part of the minimum annual charge. This redundant arrangement guarantees the maximum reability for the signal device.

The quantities gauged on the electrical panel are the battery voltage, the battery current output and the voltage of the unregulated strings before the clamping diode . A block diagram of the power conditioning system is shown in fig. 2.

The battery of the main PV system consists of 60 elements, Fiamm SMBF6 type, having an overall capacity of 600 Ah/10h and an output voltage of 120 VDC, while the emergency system emploies 6 elements Fiamm SMF 10 type, with a capacity of 500Ah/10h and an output voltage of 12 Volts.

Batteries are lead-acid stationary ones with tubular plates; their main characteristics are:

- no topping-up is required for three or more years under normal float charge conditions
- long and reliable life
- very good charge retention (self-discharge rate lower than 10%/year)
- good cycling capability

4. DETAILS OF PV USER APPLIANCES (EQUIPMENTS)

Appliances are normal marine light, with an operating range of 20 nautical miles.Light power consumption is 1000 Watts and signalling duty cicle is 1 second on and 4 seconds off.

The lamp is activated by a light sensitive switch, mean working time during the year being, obviously, 12 hours per day.

5. MONITORING SCHEME

Data acquisition system cosists of a data acquisition unit, a personal computer and a mass memory storage.

The data acquisition system is housed in a standard 19" 4U container, with rack mounting. It is composed by standardized boards having a Eurocard format of 10 cm x 16 cm. size. They perform the following functions:

- signal treatment

The current and voltage measurements are made by means of 100 mV shunts and voltage dividers. A Burr Brown ISO 100 amplifier, guarantees galvanic insulation between the photovoltaic array and the measuring system.

- analogue scanner and A/D conversion

The scanner is built to handle 32 channels, of which the system uses only 18. Each channel is supplied with overvoltage protection by using a transient suppressers of the General Semiconductor type P6KE18C.
The Analog/Digital conversion is performed using National Semiconductor devices, and an acquisition precision of 1% of the full-scale value is guaranteed.

- CPU Z80

The CPU board uses the Z80A microprocessor, a 2.5 MHz clock, an 8k ROM, a 2K RAM, and completes one acquisition cycle every two seconds, and stores in memory the following data:

Channel	Type of measurement	Unit
0	Analog test	V
1 - 3	Voltage in PV sub-arrays 1-2-3	V
4 - 6	Current " " " " "	A
7	Battery voltage	V
8	Battery charging current	A
9	Load current	A
10-12	Temperature in PV arrays 1-2-3	deg. C
13	Temperature in battery room	"
14	Outside temperature	"
15	Solar radiation on horiz.plane	kW/m2
16	Solar radiation on PV array plane	kW/m2
17	Wind speed	m/sec.
18	Wind direction	deg
19-31	Not used	

Each measurement cycle has the date and hour associated with it. Furthermore, it was felt suitable to acquire and store in memory the following statuses:

1 twilight switch operation,

2 emergency light operation,

3 command for main light to operate intermittently (to flash).

The arithmetic average. of the data thus acquired is made for each hour of acquisition, the purpose being to obtain greater precision in the data gathered, while cutting down the volume of data stored.

The system is then be able to accumulate data for four consecutive months.

The data acquisition software is written in ASSEMBLER Z80, and is stored in the ROM memory of the UA2 unit. The data processing and diagnostic software is written in BASIC A (IBM). The operative system used is the DOS 2.11. The support used consists of dual-face dual density disks, and utilizeable on the IBM PC/AT, PC/XT, Olivetti M24, and Sharp PC/7ooo personal computers, and all the IBM compatible's (6th level).

The data processing software lets the following data be computed, stored in memory, displayed, and printed out: incident solar energy; electrical power generated; losses in the cables; losses in the voltage regulators; battery efficiency; PV array efficiency. Furthermore, a summary and a listing of the data acquired and processed is generated.

The diagnostic software signals any system anomalies, and helps the operator in calibrating and setting up the analog parts of the sensors.

6. EXPERIENCE OBTAINED DURING INSTALLATION AND COMMISSIONING

Main difficulties in plant erection were those encountered during installation, material transport and electrical and mechanical erection works.

Difficulties due to the natural position of Palmaiola island have been increased by the total absence of landing point, by shallow waters, and by the steepness of the only access lane to the lighthouse.

Abovesaid problems have been overcome with an helicopter rented to transport material and a small fishing boat to transport technical personnel.

The helicopter, taking off from Cavo, in Elba island, trasported all materials in a little artificial space, opened near the lighthouse while the boat leaving from Cavo too, brought in and out, in the morning and the evening, the installing team. The team was composed by 6 workers, 1 foreman of subcontractor company (SEI-Prato) and one Pragma supervisors.

In spite of above said problems and of adverse weather conditions the whole plant was mounted in eight days and made perfectly working.

Pragma

PALMAIOLA ISLAND
PHOTOVOLTAIC
LIGHT HOUSE
BLOCK DIAGRAM

COMMESSA - JOB | IMPIANTO - UNIT

DIS. N° / DWG. No 11

Fg. - Sh./di - of: /

Rev. 0

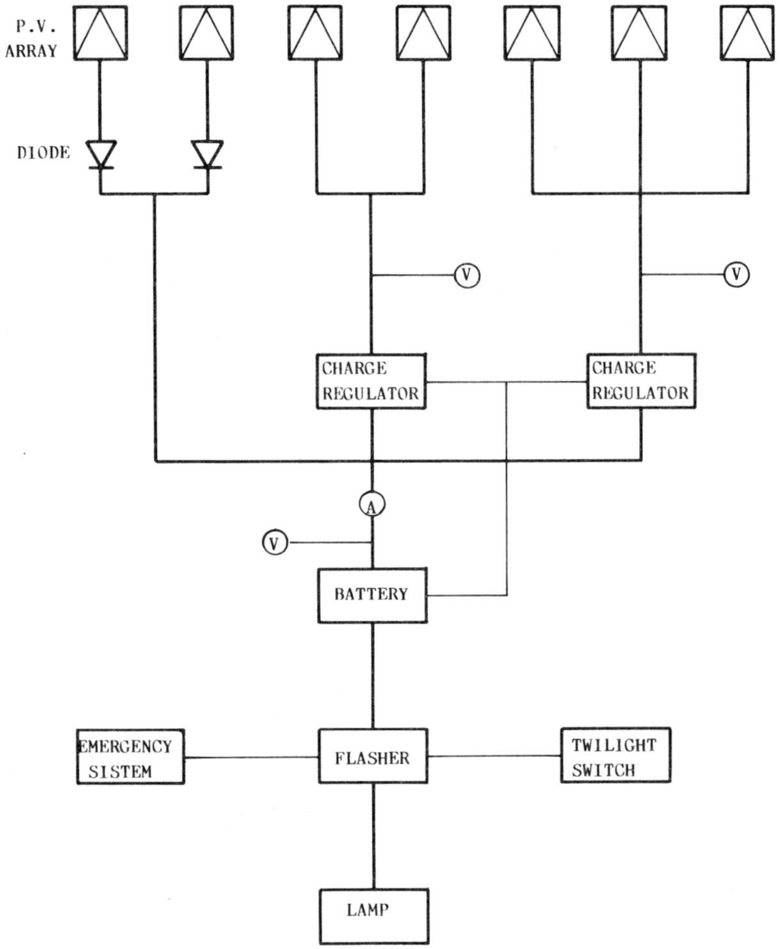

Rev.	Descrizione - Description	Comp-Prep'd	Ver-Chk'd	App-Appr'd	Data - Date
0	Emissione - Issue				

STE - A. 009

CONTRACT NO. SE 704 83 HE

PHOTOVOLTAIC INSTALLATION OF THE R. F. STATION ANTIKYTHIRA

H. KADELLA *) and D. TOUZOPOULOS **)
*) Siemens Solar GmbH, Bergisch Gladbach/FRG
**) Hellenic Telecommunications Organization (O.T.E), Athens/Greece

Summary
The objective of the project is to demonstrate that an isolated telephone, radio and television repeat station can be power supplied from a PV-plant in cooperation with an existing diesel plant. The station is located on Antikythira Island, between Crete and Peloponessus. The electric power of 20 kWp is generated by 160 solar modules. Eight Varta bloc batteries with a capacity of 13600 Ah are installed. Three separated small solar modules are foreseen for charging the common type starter batteries of the diesel generators. The PV-plant meets the energy demand of two different types of consumers. The 24 V DC consumers are the monitoring system, receiver and transmitter, fan, emergency room light, and the 220 V AC, 50 Hz consumers are antenna light, airway beacon, compressor and air-dryer, carrier telephone and voltage relais. The installation and commissioning was finished at the end of October, 1985 and the data acquisition system has been operating since November 1, 1985. An evaluation of the data of the monitoring system made by JRC-Ispra shows that since that day which is also the date of the provisory acceptance the PV-plant has been running very well.

1 INTRODUCTION

The 20 kWp-plant was installed on the Greek island Antikythira, between Crete and Peloponessus, to generate electric power for an isolated telephone, radio and television repeat station.

Besides the PV-array, there are three small PV modules mounted on the roof of the station building for loading the starter batteries of the diesel engines.

The construction of this photovoltaic plant was partly sponsored by the European Community. During the planning of the plant care was taken that it was tailored to and integrated into existing components such as diesel aggregates and auxiliary systems.

The design of plant and equipment was orientated towards keeping losses and need for maintenance as low as possible and achieving high reliability.

The battery has a storage capacity of 13600 Ah for a rated voltage of 24 V. Its purpose is to balance the fluctuations between the energy supply, which is determined by diurnal variation and by clouds, and the energy demand of the radio relay station including the auxiliary systems. The battery was designed for 9 to 10 sunless days.

The running measuring programme is delivering the different energy flows and the characteristic meteorological values. The measured values averaged over one hour are stored on a magnetic tape cassette (Fig. I).

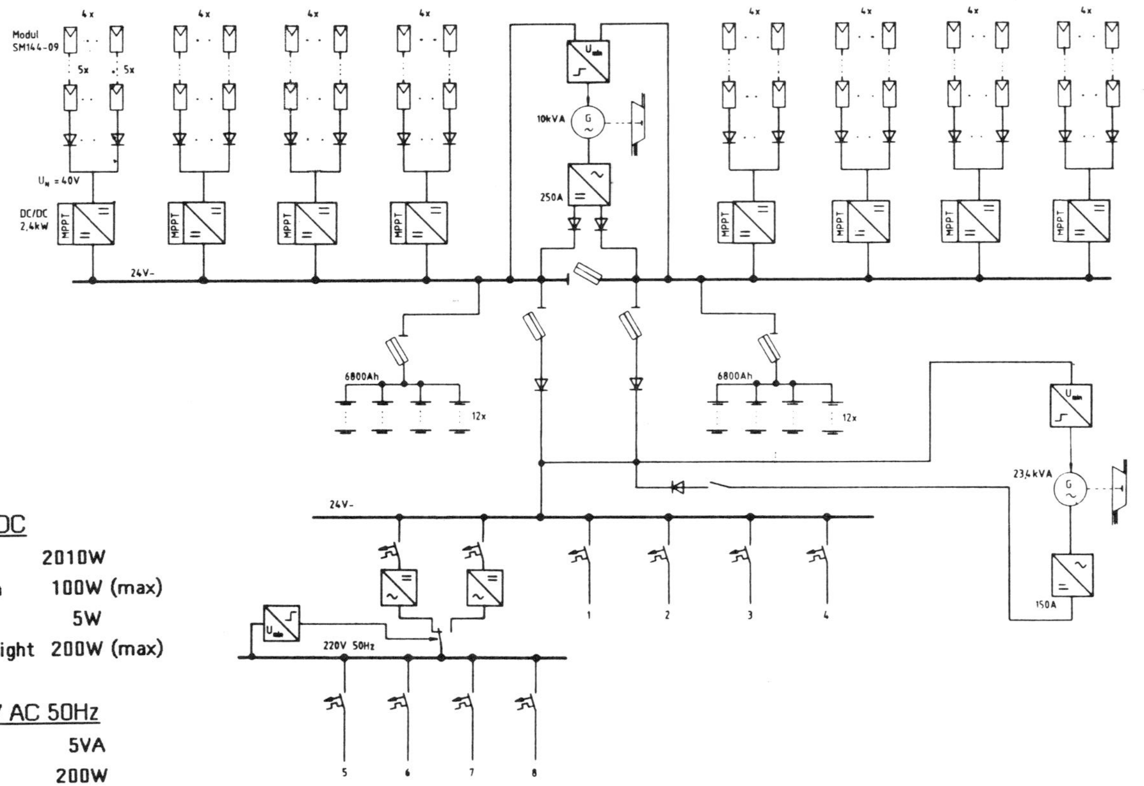

CONSUMERS 24V DC

1	Transmitter	2010W
2	Monitoring system	100W (max)
3	Fan	5W
4	Emergency room light	200W (max)

CONSUMERS 220V AC 50Hz

1	Voltage relais	5VA
2	Antenna light	200W
3	Air–dryer	260VA
4	Carrier telephony	65VA

Fig. I: One-line diagram power supply

2 PHOTOVOLTAIC ARRAY

The solar generator with a rated power of 20 kWp consists of 160 Siemens Solar Modules SM 144-09. Five modules are connected in series to form a branch. Four branches are switched in parallel to form an array unit which is allocated to a DC/DC converter (2.4 kW).

Eight parallelly connected array units form the whole solar generator. The needed area for the modules is approximately 1000 m^2, the tilt angle is 60°.

Technical data of the solar modules SM 144-09 under standard conditions (cell temperature = 25 °C; AM = 1.5; Insolation = 1000 W/m^2):

- Short-circuit current I_k	16 A
- Open-circuit voltage U_L	9 V
- Output power P_{max}	120 W
- Rated voltage U_N	8 V
- Dimensions in mm	1470 x 1020 x 80
- Weight	27 kg

3 POWER CONDITIONING

The plant consists of two independent supply lines which feed into a 24 V DC consumer busbar. Each line comprises 4 DC/DC regulators. The 8 DC/DC converters are equipped with autarkic MPP controllers (MPP - Maximum Power Point) and superimposed charging voltage limiting controllers, the charging characteristic curve which meets the battery charging regulations.

The plant is designed without fuses. All protective functions have been realized with monitored automatic power circuit breakers or cut-outs. The two 24 V feeder bars are coupled with an automatic power circuit breaker. In the case of an accident, these busbars are decoupled via the adjustable over current or short-circuit trip. One battery with 6,800 Ah is allocated to each feeder bar. In addition, both feeder bars are supported by a 10 kVA diesel via a charging rectifier (250 A). The automatic start-up of the diesel is initiated via an undervoltage trip. The operating time is set via a time relay in accordance with the necessary recharging of the battery.

The 24 V consumer busbar is fed via 2 diodes decoupled automatic power circuit breakers.

A 23 kVA diesel is available for the further support of the consumer busbar. Start-up of this diesel is initiated via an adjusted undervoltage relay which is staggered with the 10 kVA diesel. The operating time is set via a time relay for a night cycle.

The 220 V AC busbar is fed from the 24 V consumer busbar via an inverter (1,000 VA). Further auxiliary systems are supplied from this bar. If the inverter fails, the second hot stand-by inverter will feed the 220 V AC busbar with an interruption of less than 500 ms.

All disturbances are compiled in a group alarm. This group alarm or special individual alarms are available for telecommunication (telephone).

The overall plant is provided with multi-stage protection against atmospheric overvoltages. Lightning protection masts are distributed in such a way in the solar field that a direct lightning strike will be intercepted by the masts and conveyed to the earthing network system.Surge arrestors are installed in the input of the power distribution cubicles (upstream of the DC/DC converter) as protection against induced voltages. Varistors and capacitors are installed in the input of each DC/DC converter as protection against induced voltages. Varistors and capacitors are installed in the input of each DC/DC converter. Each of the two 24 V feeder bars is equipped with 2 non-linear resistance arrestors for + and - pole.

The 24 V consumer busbar has a non-linear resistance arrestor at the - pole. The + pole is earthed (due to the design as a result of the transmitter).

All decoupling diodes are protected by varistors.

4 INSTALLATION AND COMMISSIONING

The PV plant was installed under aggravating circumstances within approximately one year. Poor site supporting facilities, insufficient loading conditions at the harbour and bumpy narrow gravel roads made the transport of the hardware very difficult.

200 kg dynamite were needed for levelling the ground and laying the foundations of the module supporting structures. Since the space available in the station is needed for the maintenance personnel a special housing had to be built for the batteries.

The commissioning including the training of the O.T.E. personnel on site was finished according to the "Delivery Report" of October 28, 1985.

During the commissioning, it was found out that one PV-module was defect. This defect module could not be exchanged because the two spare modules on site were destroyed during the blastings due to improper storage of the modules.

5 OPERATION AND MAINTENANCE

Operation

The plant is operating automatically after commissioning and functional testing.

Maintenance

- Battery

  The application of so-called recombinators increased the maintenance intervals to ≥ 3 years. However, it must be pointed out that regular inspection of the functional safety of the battery (acid density measurement, cell voltage) is absolutely essential, also in the case of "maintenance-free" batteries.

- Switchplant

  The switchplant is maintenance-free. However, a functional test of the diesel aggregate cut-in in conjunction with the rectifiers should be performed annually.

- Diesel Aggregates

  In order to retain the operability of the diesel aggregates, these are to be operated with at least 70 % of their rated power for approximately 2 h in a monthly cycle. Diesel fuel should be stockpiled so that the supply is renewed annually.

Experience

On December 23, 1985, we received the message from the OTE that the 10 kVA diesel could not start because the old starter battery was out of order.

- last month the auxiliary diesel engine was activated twice, due to voltage drop of the batteries perhaps of the extremely hard winter time this year
- the data acquisition system sometimes ceased working
- the existing old transmitter has recently got operational troubles due to difficulties with its voltage stabilization.

The problems which recently occurred could be solved by the O.T.E. personnel which was trained by Interatom twice.

This PV-trained personnel is no longer available to look after the plant because of the high fluctuation at O.T.E.

Despite these recently arisen problems caused by the old O.T.E.

equipment of the R. F. station an IRC-Ispra evaluation of the data received from the monitoring system indicates that the PV-plant has been running very well since the provisory acceptance at the end of October 1985.

6 MONITORING

6.1 Data Acquisition System Hardware

The acquisition unit consists of the control computer HP9915A with monitor and keyboard, a digital voltmeter with multiplexer input cards HP3421A, a magnetic tape cartridge unit CDS 2.50 manufactured by Meß- und System-Technik GmbH, and a DIN A4 Matrix Printer HP82906A. All component parts are connected to each other via the Hewlett-Packard Interface Bus (HP-IB). With the exception of the printer they are positioned in a medium height of 19", cabinet equipped with viewing door. The matrix printer is positioned on top of the cabinet to ensure ease of operation and paper loading.

The complete unit is supplied from the 24 V busbar via a common connection. The supply cable and measuring cable are connected from the rear to a terminal strip positioned in the cabinet.

The station is operated from a separate keyboard which is installed underneath or beside the monitor (Fig.II).

Fig.II: Configuration of the Data Acquisition System

6.2 Acquisition Program

The following variables are measured:

- Global radiation
- Insolation on panel level
- Ambient temperature
- Module temperature
- Output energy PV field
- Output energy DC/DC converter
- Input energy battery
- Output energy battery
- Energy to the consumer
- Energy from the diesel generators

The acquisition program initiates a scan of all connected measuring channels and accepts the signals with time statement in a rough data field. The rough data are then subjected to a measuring range check. If a value is outside the possible operating range, an announcement is made via the printer. The rough data are then converted to the necessary physical values (temperatures, insolation powers, electrical powers), using a pre-specified conversion matrix. Thereby some measuring signals are used for different derived values (e. g. battery input and output power).As far as the diesel power is concerned, a differentiation is made as to whether the feed is with or without plant demand (digital inputs). The measured values of the last full hour or those which have occurred from the starting time-point are averaged and intermediately stored in a mean value list. For the full hour the mean power values are multiplied with the acquisition period (generally 1 hour). Thereby one obtains the energy values (EV) from the mean values (MV). The mean temperatures and the energy values are recorded on the magnetic cartridge with date and time.

The program allows to check and, if necessary, to change the time and date. In addition, it is also possible to continuously display one or more measured values during acquisition.

List of Mean Values:
MV1: Ambient temperature
MV2: Module temperature
MV3: Global radiation
MV4: Radiation on panel level
MV5: Power PV fields
MV6: Power DC/DC converter
MV7: Input power battery
MV8: Output power battery
MV9: Consumer power
MV10: Power diesel 1 with demand
MV11: Power diesel 1 without demand
MV12: Power diesel 2 with demand
MV13: Power diesel 2 without demand

The hourly energy values and mean temperature values are recorded on the magnetic cartridge at hourly or longer intervals.

PV PUMPING FOR IRRIGATION
PROJECT NO. SE/215/83HE

Name and address of contractor	HELLENIC INDUSTRIAL DEVELOPMENT BANK SA 18, Panepistimiou street GR 106 72 ATHENS, GREECE
Site, country	Menetes, Karpathos island, Greece
Rated peak power of PV-array	10 KW
Starting date of the project	7.1.1984
Starting date of 2 years monitoring campaign	1.10.1986
Name and address of the manufacturers of the main components	BP Solar system Ltd, UK: PV array Pleuger of Greece : water pump AEG Telefunken : Inverter 10 KVA Athens University : Electronic control, Inverter 600 VA and Data logging system
Contractor's project leader	1. Prof. J. Avaritsiotis, Technical University of Athens, Department of Electrical Engineering. 2. E. Vazeos, BP of Greece, Solar Dept.

SUMMARY

The project of "photovoltaic pumping for irrigation" is in progress at Karpathos island, Greece. Besides technical experience to be gained from the project, development of a remote area will be promoted.

The inclusion of a diesel generator as an emergency back-up, proved to be necessary due to the necessity of uninterrupted water supply since the erection of commercial greenhouses in the area is envisaged.

The main problem encountered during the construction period is the difficulties of transportation to the island.

The system was designed to meet the irrigation needs of the region in the most satisfactory manner as well as to achieve the best possible utilization coefficient for the photovoltaic generator.

1. INTRODUCTION

Hellenic Industrial Development Bank S.A. (ETBA S.A.) in

co-operation with Athens University and B.P. of Greece has already installed a PV solar pumping system for irrigation at Karpathos island in Greece. The solar array with a total peak power of 10 KW feeds a submersible centrifugal pump through an 10 KVA DC/AC inverter with constant V/f output.

Pumped water is stored in two tanks, already existing. The operation of the system is controlled automatically.

The monitoring system provides a comprehensive range of data for the operation of the system and subsequent data analysis.

The equipment are protected inside a small building which has been erected above the well's opening.

2. PHOTOVOLTAIC ARRAY

The solar array field consists of 309 solar modules, type BP A 1233, manufactured by BP SOLAR SYSTEMS LTD, with a total peak power of 10 KW.

The photovoltaic generator is divided into two sections:

The first section (power 8,850 Wp) feeds a submerged centrifugal pump through an inverter (10 KVA DC/AC). The second section (power 1,150 Wp) constitutes part of an autonomous photovoltaic system which includes batteries of 600 Ah, and a 60 V DC/220 V AC inverter; it provides the necessary electric power for both the data logging system which uses a micro-computer and the automatic control system.

The modules of the main system are arranged in sections each containing 7 modules connected in series. Three adjacent sections are connected in series to form a sub-array of 21 modules, with a nominal terminal voltage of 330 V DC. The sub-array for the auxiliary system consists of 5 modules connected in series. The output of each of the 17 sub-arrays (14 for the main and 3 for the auxiliary system), is transferred to the control cabinet via a 2 core NYY cable. The layout of the array field is shown in Fig. 1.

Each section is supported by a galvanised steel framework. The design of the support structure is shown in Fig. 2. The inclination of the modules is 35° towards south. The ground where the solar array is mounted is rocky and rough. This has been taken into consideration in the design of the supports as well as in the type of foundation that was selected. The metallic structure is electrically earthed as a protection against lightning.

3. POWER CONDITIONING

3.1. Main System

The pump is powered by the solar array through a 3-phase inverter with sinusoidal output current and variable voltage amplitude and frequency. The inverter is the 10 KVA Solarverter (G) of AEG which is applied in solar systems without a battery. It is also equipped with a MPPT (Maximum Power Point Tracker).

The ratio of the output voltage to the output frequency is maintained at a constant value for different frequencies, so that the load of the pump water will be always operated with a constant magnetic flux and can be loaded with its rated momentum at all rotational speeds, up to the rated rotational speed. At the lowest frequencies the voltage is increased in order to compensate the ohmic drops in the stator of the motor.

The technical specifications of the above mentioned inverter are given in the following table I.

TABLE I

Rated output power	10 KVA
Rated output current	25 A
Output voltage	0...0.75 U
maximum (at U_{in} = 345 V)	255 V
Output frequency	
minimum/maximum	1/500 Hz
standard frequency range	1...60 Hz
Rated input voltage U	300 V DC
U_{in} minimum/maximum	200/250 V DC
Efficiency in the rated point	95%
Environmental temperature	0...40°C
Type of cooling	self cooling with air
Type of protection	IP 20
Weight	approx. 50 kg
Dimensions in mm	480 X 754 X 337

3.2. Auxiliary system

The auxiliary PV system has constant voltage output. A bank of batteries is used with a total capacity of 600 AH. A solid state battery charge regulator, sensing the battery voltage and the electrolyte temperature safeguards the batteries against overcharging.

The auxiliary system provides AC power through a 600 VA DC/AC inverter constructed by Athens University. In Fig. 3 the block diagram of the system is shown.

4. SPECIFICATIONS AND OPERATION OF THE SYSTEM

The level of the water in the well varies with the pumping flow rate, from 8.1 m down to 69.5 m (depth of the well). As an example, after 8 hours of pumping at a rate of $8m^3/h$, the level of the water in the well dropped to 32.4 m from the opening of the borehole. The pump is placed at a depth of 50 m.

The pump is of submersible, centrifugal type coupled to a watertight 3-phase motor with a rated power of 6.3 KW at 380 V. The manufacturer of the pump (type N65-8) is Pleuger. In Fig. 4 the characteristics of the pump for various speeds are shown. In the same diagram the characteristics of the hydraulic network towards both tanks have been superimposed.

The water will be pumped either to a tank positioned 12 m below and 1,550 m far from the well opening or to a second one positioned 27 m above and 264 m far from the same point. The changeover from one tank to the other will be done automatically according to the power output of the inverter, i.e. if the inverter output exceeds a predetermined threshold value, the water is pumped to the upper tank. It is worth noting that from simulation studies, performed independently by Athens University and BP, the total volume of the water pumped to both tanks depends dramatically on the threshold level.

The actual performance of the P/V water pumping system in Karpathos is in agreement with the results of the simulation study that was carried out during the design stages. For the period of November the amount of water pumped from the well was in the order of magnitude calculated in the simulation study.

Figures 5 to 17 show the amount of water pumped during November 1986 in relation to the incident solar energy and the relation of several parameters of the station for two days of the same month.

5. INSTALLATION OF THE SYSTEM

The installation of the system is completed. The most serious problem encountered, so far, is the difficulty in transportation to the location, (specially in winter time), due to the remoteness of the area. Monitoring data are not available for a period of 4 months (Dec. 1986-Mar. 1987) due to the breakdown of the auxilliary inverter that powers the monitoring system and simultaneous problems to the cable feeding the water of the pump (2 times).

6. MONITORING

The parameters to be monitored are:

- insolation (horizontal)
- insolation (at the surface of PV array)
- Ambient temperature
- Wind velocity
- Cell working temperature
- power o/p from the PV array
- power o/p from the 10 KVA inverter
- water flow
- water level in the well, and finally various events i.e. switching of water flow, functioning of the diesel generator, etc.

The data recording frequency is 3 min. with integration during the hour.

The data logging is performed with the aid of a personal computer (fully IBM PC compatible) and signal conditioning cards (μMAC-4000) manufactured by Analog Devices.

The data are recorded on floppy disks (360 KB DS/DD) which are sent to Athens every fortnight. A copy of the data on floppy disk is sent to the E.E.C.

7. AIM OF THE PROJECT

The benefits arising from the use of such a photovoltaic pumping system are:

a. direct
   - under-cover cultivation, using the water which will be pumped for irrigation
   - incentive for the establishement of small hotels in the region
   - development of photovoltaic technology in Greece
b. indirect
   - development of technology pertinent to intensive under-cover cultivation
   - creation of technological infrastructure.

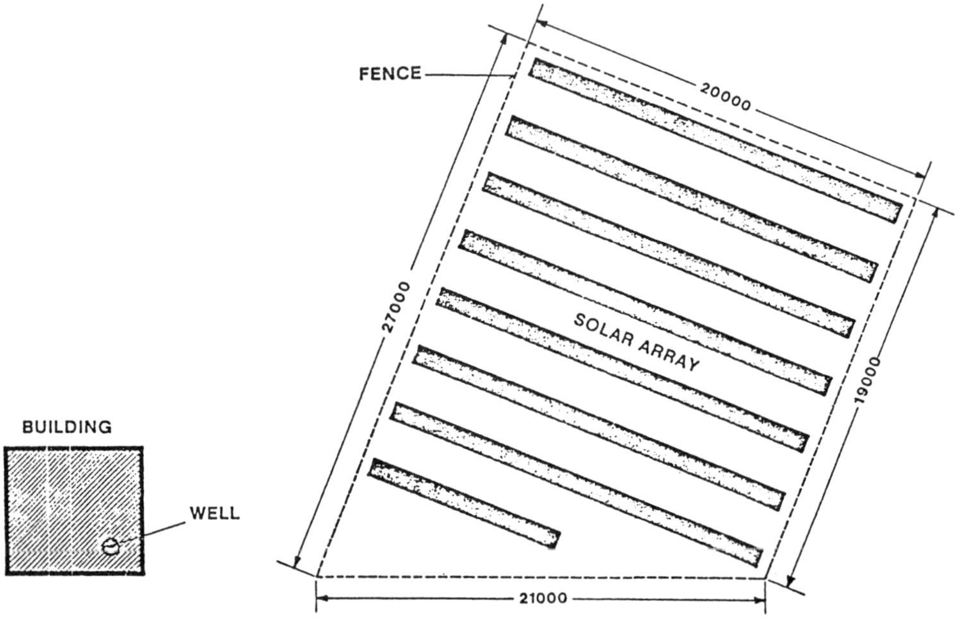

Fig. 1: ARRAY FIELD LAYOUT

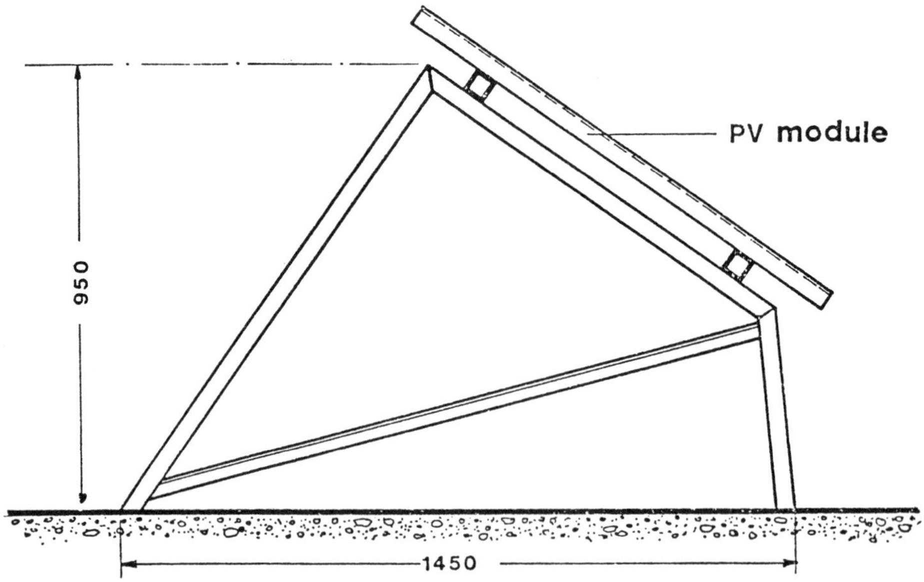

Fig. 2: SUPPORT STRUCTURE

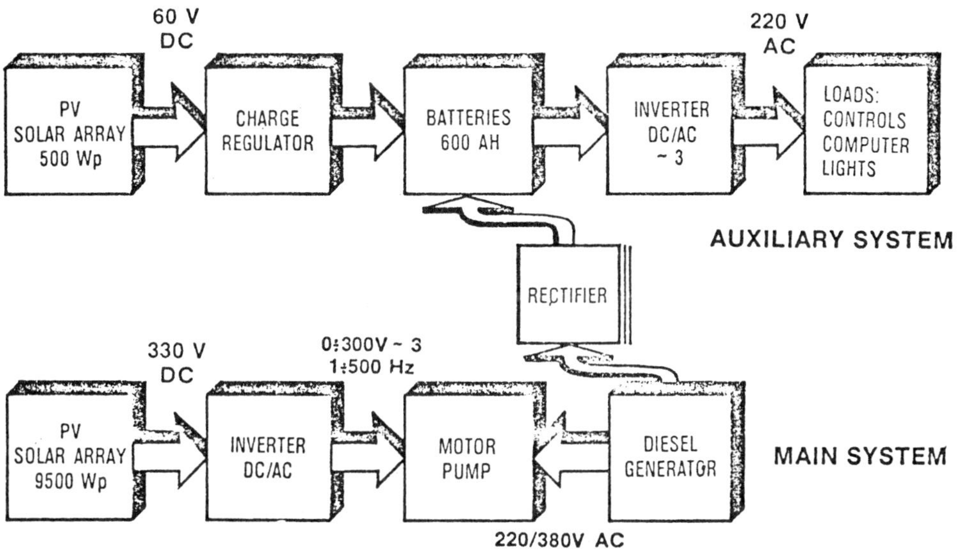

Fig. 3: BLOCK DIAGRAM.

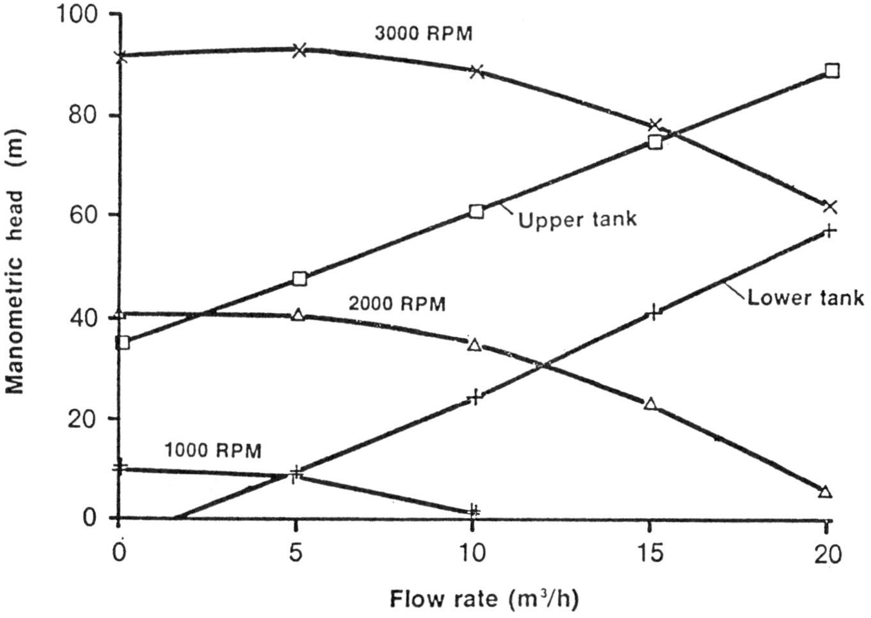

Fig.4: CHARACTERISTIC CURVES OF PUMP AND NETWORK

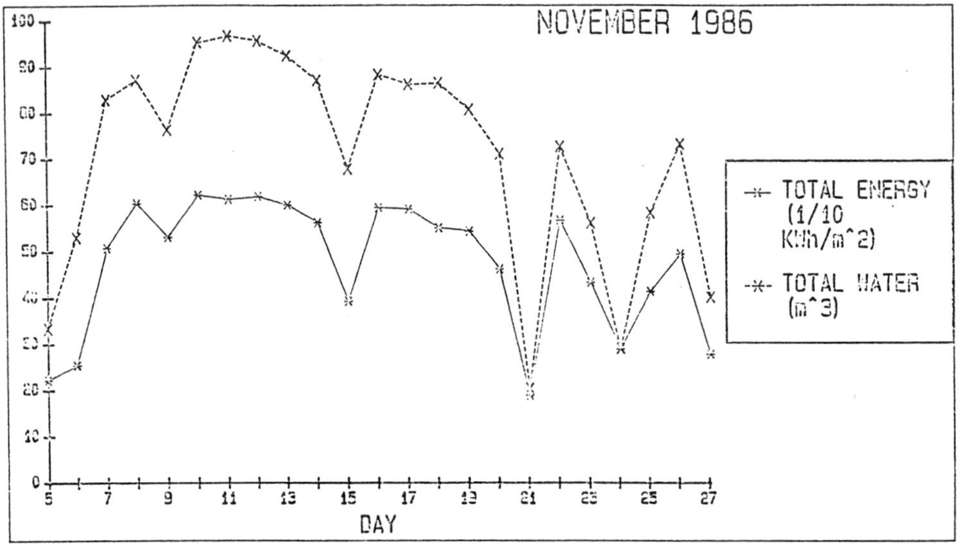

Figure 5

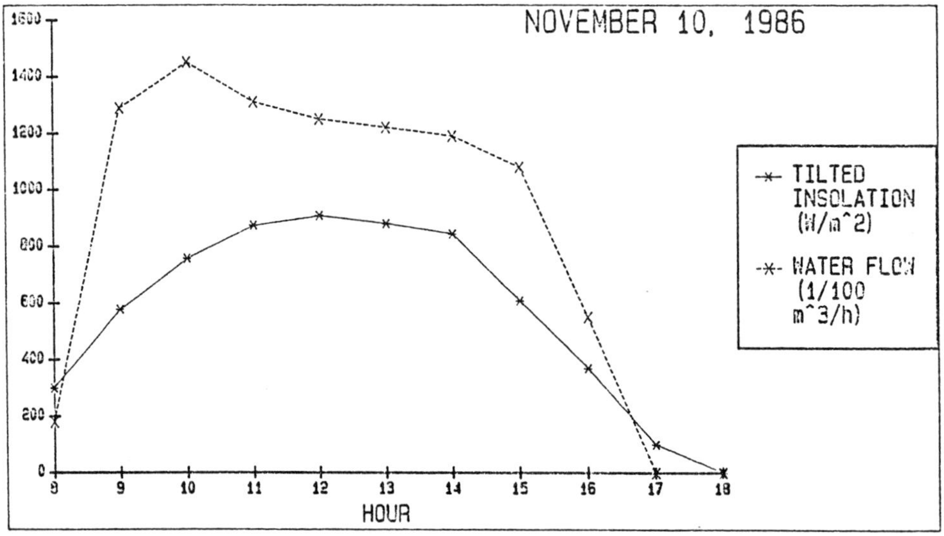

Figure 6

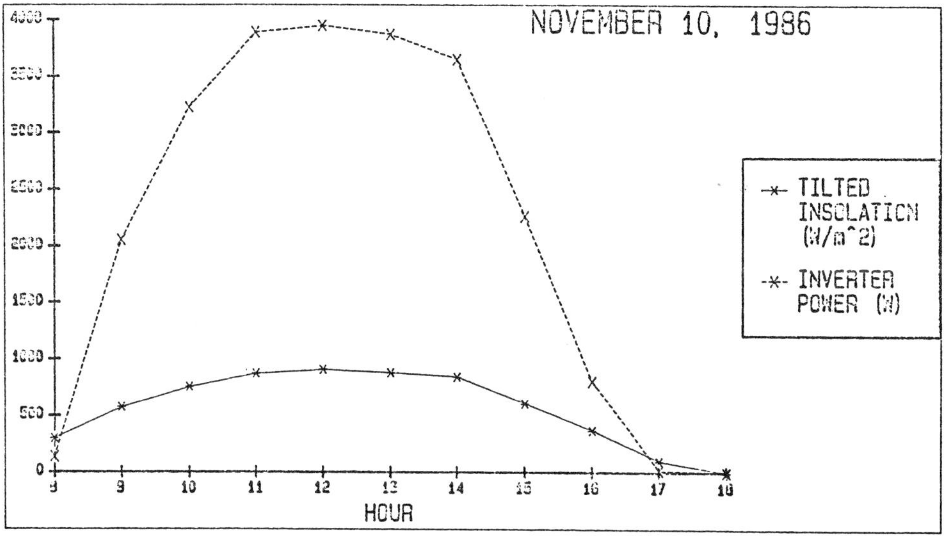

Figure 7

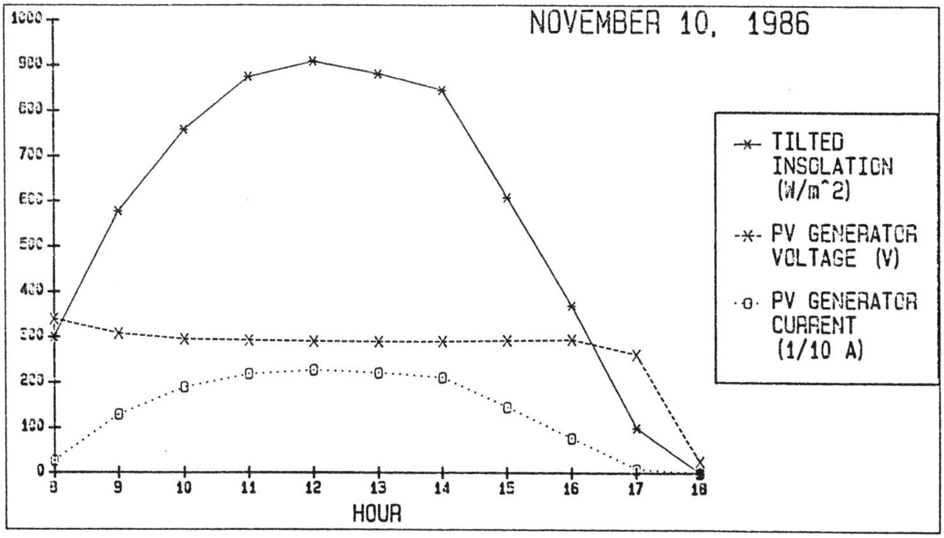

Figure 8

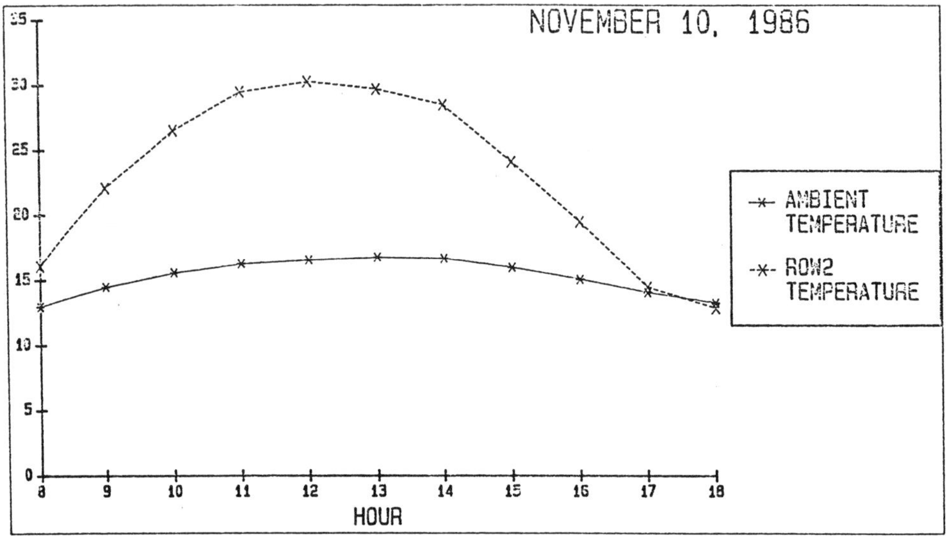

Figure 9

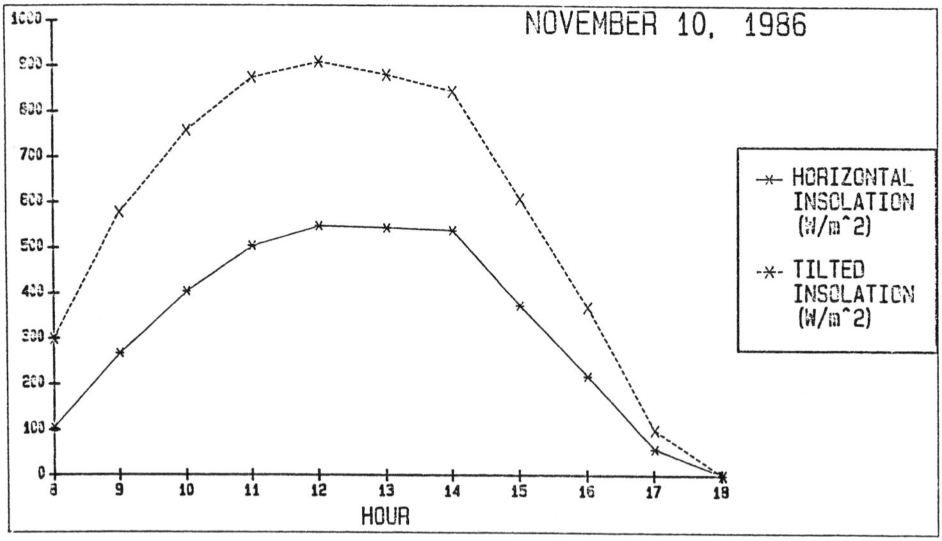

Figure 10

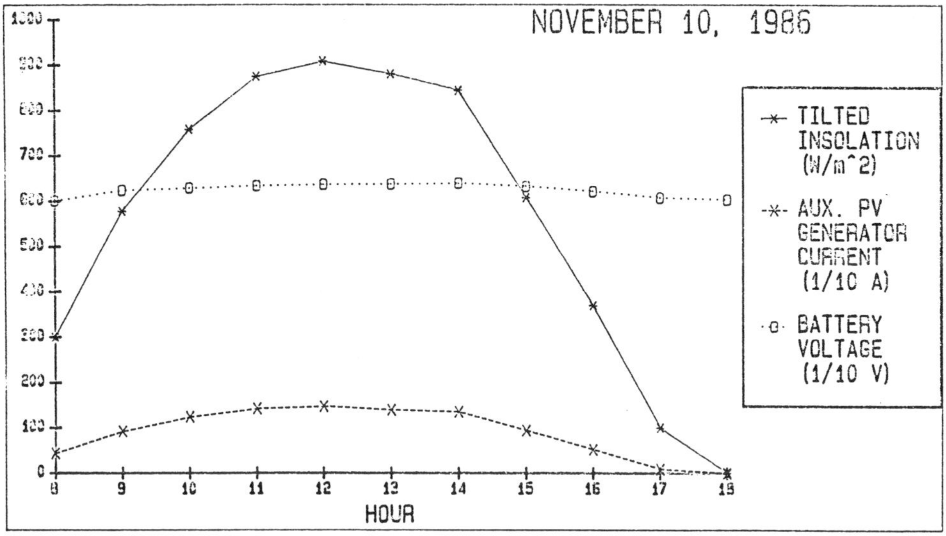

Figure 11

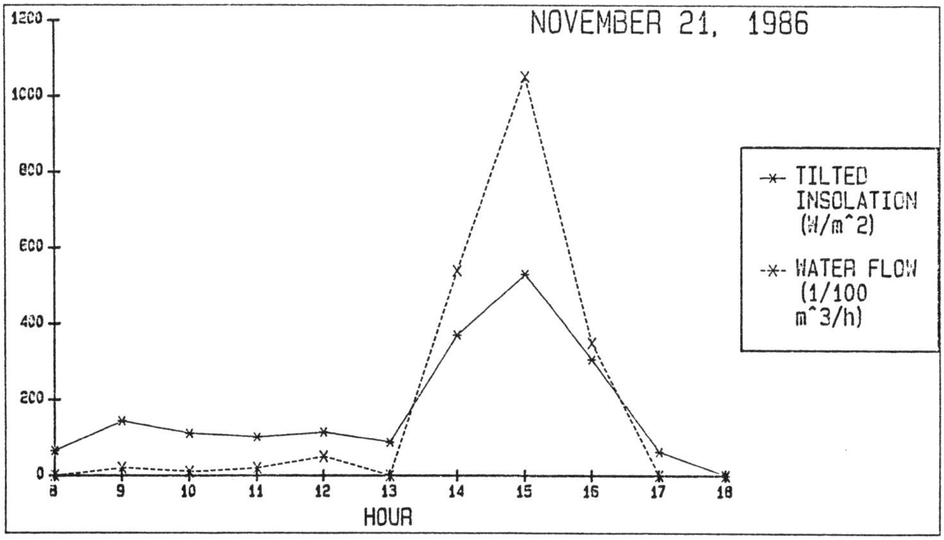

Figure 12

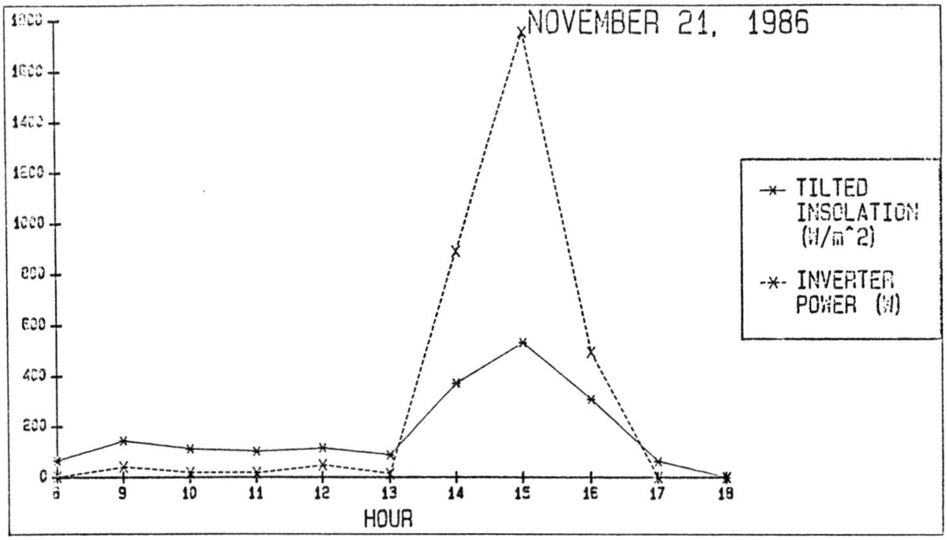

Figure 13

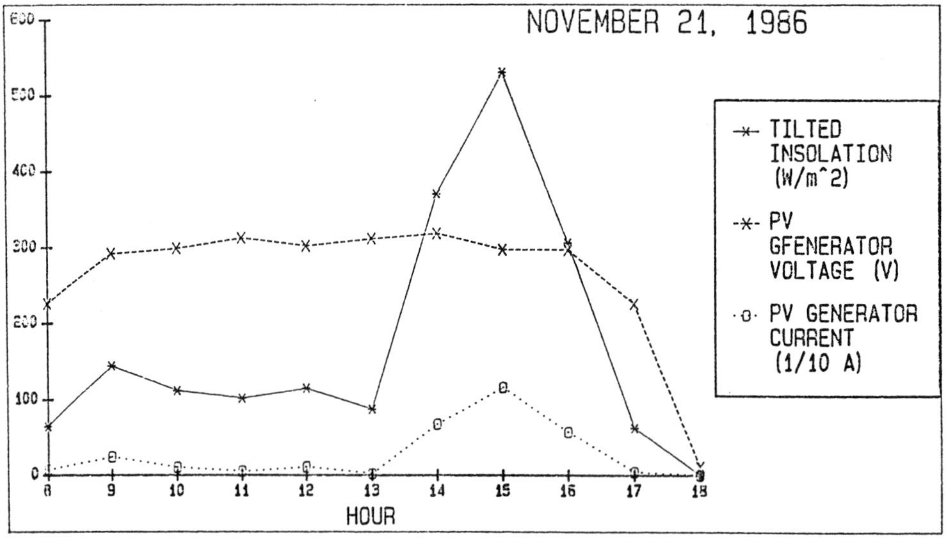

Figure 14

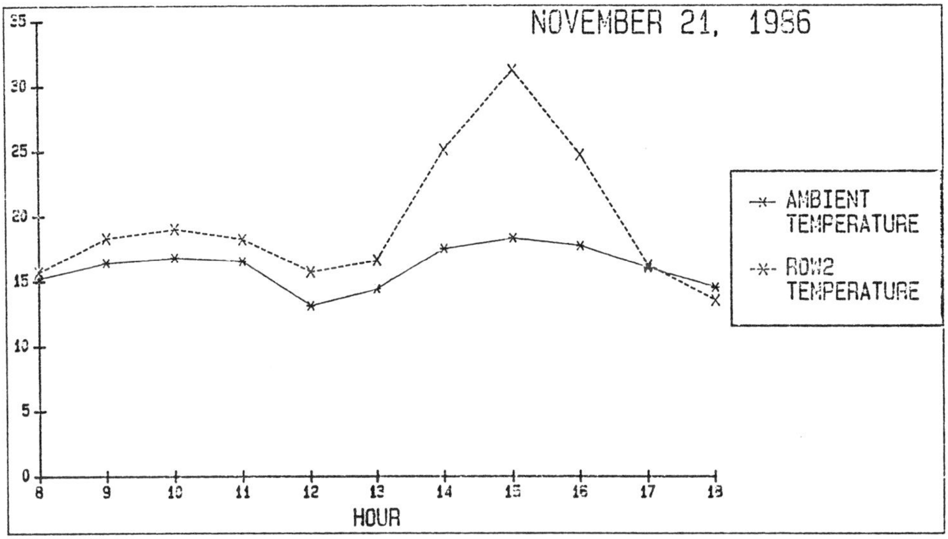

Figure 15

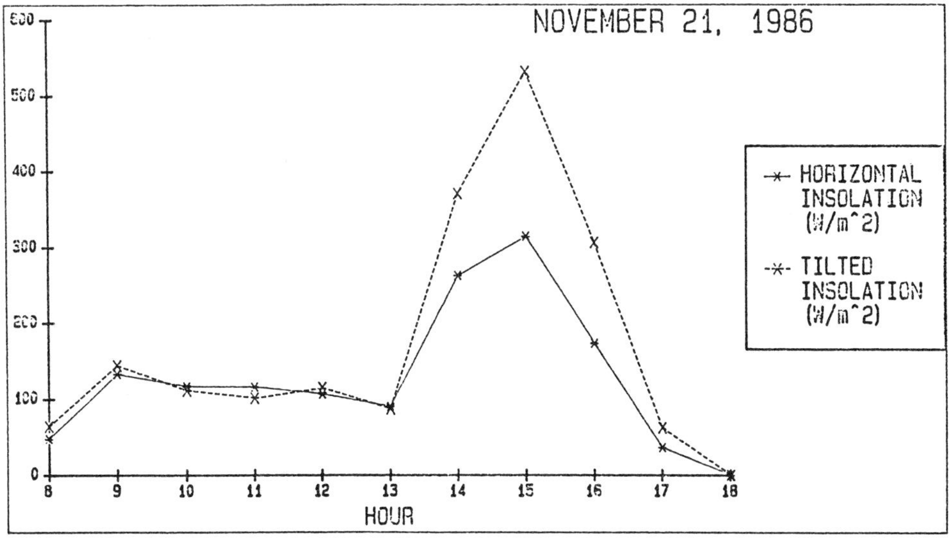

Figure 16

CONTRACT NR SE/123/83

PHOTOVOLTAIC SOLAR CURRENT GENERATOR TO SUPPLY AN EARLY WARNING SYSTEM FOR FORESTAL FIRE

F. FONZI
Technical Director of Italenergie S.p.A.

Summary

Photovoltaic solar system with microprocessor central unit, equipped with particular sensors, allows the monitoring of the meteorological and climatic physical quantities, even of catastrofic nature.
The system is able to:
1) detect quickly forest fires in ten kilometers range and operate an early warning system;
2) forecast frost about 14-12 h in advance with 95% certainty;
3) forecast river floadings in correlation with rain fall intensity.
The system includes the following essential parts:
- photovoltaic current generator
- electrical energy storage batteries
- electronic detection unit
- data recording and transmission
- data elaboration unit
- monitor and control computer
The system represents one unit of an early warning system network to protect alarge territory.

1. INTRODUCTION

The present photovoltaic solar system for early warnings (forestal fire, frost and inondation) to protect the potential of biomasses as energy sources is composed of the following parts:
- photovoltaic power unit: the power unit is represented in Fig. 1. It is composed of 12 photovoltaic modules with a battery set. An electronic charge control device protects the batteries;
- data acquisition unit: the unit has an infrared sensor to localize spots with temperature above 150°C, optical lens to check smoke emissions in the controlled area, electromechanical device to check the water level, more sensors for frost prediction;
- directing unit: the data acquisition unit can rotate to watch the landscape all around and can perform also azimutal mouvement;
- steering computer unit: this unit provides the control and regulates all the functions of the system: data analysis, coordinate calculations, fire

extension, water level control, frost prediction calculation, alarm time, steering of the directing motors, regulation of the view field of the optical unit, diagnostic of the whole equipment with indication of failures;

- transmitter unit: the collected data are transmitted by this unit using transmission frequency of 445,900 MHz. The optical image is transmitted with a frequency of 23 GHz. The Figg. II, III and IV give further information on the unit.

2. PHOTOVOLTAIC ARRAY

Technical data

The photovoltaic array is composed by 12 Modules of Helios Technology, type H 436/1 with 396 W_p, 24 Vdc.

The main characteristics of the Helios Technology type H 436/1 are:

- peak power : 33 W_p
- short circuit current : 2,5 Amps
- open circuit voltage : 19,5 Volts
- voltage by maximal power : 15,5 Volts
- current by maximal power : 2,13 Amps
- insulation : 2000 V
- execution : tempered glass
- number of cell per module : 36
- type of cells : monocrystalline
- dimensions : 1041 x 406 x 47 mm
- weight : 5,6 kg

For the measures in the field there is the need to separate the early warning system for forestal fire, which remains in Sulmona area, from the early warning system for flooding and frost, which is installed by Pescara river, near the town of Pescara.

For the second system they use a photovoltaic array composed by 2 Siemens modules, type 140-36 with 66 W_p, 24 V_{dc}.

The main characteristics of the Siemens modules type 140-36 are:

- peak power : 33 W_p
- short circuit current : 2,1 Amps
- open circuit voltage : 21,2 Volts
- output current (with V_a 16,4 v) : 2,01 Amps
- insulation : 2000 V
- execution : tempered glass
- number of cells per module : 36
- type of cells : monocrystalline
- dimensions : 1055 x 428 x 47 mm
- weight : 9,3 Kg

Economical data

a) Photovoltaic current generator

- peak power : 396 W_p
- yearly produced energy : 289.080 Wh/year
- costs of investment:
  - . photovoltaic array : Lit. 14.300/Wp
  - . battery charging regulator : " 2.020/Wp

. batteries	:	Lit. 5.051/Wp
. structures	:	" 4.798/Wp
. cables, wiring and others	:	" 1.610/Wp
Total	:	Lit. 27.779/Wp
		ECU 18,52/Wp
- Total cost of the investment	:	396 W_p x 27.779 Lit/W_p =
	=	Lit. 11.000.000
	=	ECU 7.333
Amortisation period	:	10 years
Yearly interest	:	15%
Yearly cost of the investment	:	Lit. 2.192.000/year
	=	ECU 1.461,33/year

- Cost of invest. for energy unit prod:

Lit. 2.192.000/year : 289.080 Wh/year = Lit. 7,58/Wh

ECU 1.461,33/ year : 289.080 Wh/year = ECU 0,005/Wh

- Yearly operation costs	:	
. cost of investment	:	Lit. 2.191.000/year
. operation materials	:	" 20.000/y
. maintenance and repairs	:	" 100.000/y
. labor cost	:	" 200.000/y
. diary cost	:	" 160.000/y
. travel costs	:	" 83.200/y
		Lit. 2.754.200/year
	=	ECU 1.836/year

- Total operation cost for energy unit produced :

Lit. 2.754.200/y : 289.080 Wh/y = Lit. 9,51/ Wh

ECU 1.836/y : 289.080 Wh/y = ECU 0,006/Wh

3. POWER CONDITIONING

The Fig. V shows how the modules and the system have been cabled. In all cases particular care has been devoted to the connections to avoid energy losses, voltage drops, induced voltage. Protection have been applied to avoid hot-spots and reverse current by using by-pass and blok diodes. The modules have been grounded. Back-up batteries of adequate size assure an uninterrupted work of the system 24 hours on 24 hours. The absolute lack of sun for 8 days have been taken as base in dimensioning the battery. A battery charging and descharging regulator protects the batteries. The regulator operates a switch to disconnect the load if batteries charge has sunk too much. The lead-acid batteries are Varta, type VB 2306:

- voltage	:	24 Vcc
- capacity	:	300 Ah
- sunless days	:	8
- element voltage	:	2 Volts/elem.
- short circuit current	:	4,5 KA
- acid density	:	1,24 kg./lt.
- pole number	:	2
- pole diameter	:	30 mm

- weight with liquid	:	31,20 kg
- quantity of liquid	:	8,62 kg
- self-descharging %	:	< 3%

4. SPECIFICATION AND OPERATION OF THE SYSTEM

. Energy absorption of the system	:	660 Wh/day
. Energy absorbed by the motors	:	300 Wh/day
. Energy absorbed by manag. electronics	:	160 Wh/day
. Energy absorbed by the telecamera	:	200 Wh/day
. Temperature sensor	:	linearized integrated circuit
. Humidity sensor	:	capacitive thin film
. Pressure sensor	:	piezoresistance integrated circuit
. Rain fall sensor	:	digital reading with siphone outlet
. Wind speed sensor	:	optoelectronic reading
. Wind direction	:	potentiometer
. Radiation sensor	:	thermophile
. Water level sensor	:	electronic with potentiometer
. Ground temperature sensor	:	linearized integrated circuit in a stainless cover
. Humidity of the ground sensor	:	conductivity measure
. Fire detection sensor	:	infra-red telecamera
. Smoke detection sensor	:	optical telecamera
. Data acquisition and elaboration	:	CMOS microprocessor
. Data transmission	:	UHF 445,900 MHz
. Image transmission	:	radio bridge 23 GH

a) Fire alarm

The system has been installed in an adequate position so to allow the maximal exploration of the controlled area taking care to have minimal incidence angles. The system is calibrated to recognize hot spot > 500°C. The sensitivity soil is > 150°C to recognize false data, due to reflections (above all of sunshine). The data collection happens in two near line spectral regions, both situated in the transparence openings of the atmosphere.

The computer elaborates the difference of radiation intensity between the two areas of the spectrum.

The software related to the system has a standard part and a part which has to be done in dependence on the area to be set under control. This last part is realized on the field through a special service software.

b) Frost alarm

The frosts,which occur by the fall of the temperature below 0°C, are originated by two distinct natural phenomena:

- for the advent of cold air masses (caused by the movements impressed by the distribution of the pressure fields on large scale) coming from the artic circle;
- for a progressive drop of soil temperature and of the overhead airmasses, caused by the transfer of the heat,stored during the day,of the solar radiation and that cannot be compensated in night time.

Starting from the analytic description of these processes, it is possible, when one has all data on hand, to calculate with a certain approximation the thermic nocturnal drop. For particular purposes to forecast the probability of occurance of the frost one can reply upon a simplified method of the following type.

In a cartesian graph are reported on the ordinate the product of the relative humidity times the temperature at sunset and on the abscissa the value of the temperature of the dew point for the same period. Such a computation delimits the risk area for the frosts.

When the above mentioned values result to be equal or lower than 40 at the ordinate and 0 at the abscissa, one can verify the frost phenomenon with a probability of about 90-95%.

c) Flooding alarm

A mechanical sensor allows to control the increasing and decreasing level of the water with an accurancy of 20 mm. The mechanical mouvement is reported to an electronic device, which gives the measure of the level. The measurements are transmitted by 445,900 MHz to the computer.

The rain falls are measured with an accurancy of 1 mm. The correlation between the two magnitudes allows the forecas of possible flooding. The forecast has an higher accurancy and reliability if one installs more rain fall measurements spread in a wide region, where the rain waters flow together to the same river.

5. INSTALLATION AND COMMISSIONING

According to the project they have built one unit of early warning for forestal fire and two units of early warning for flooding and frost.

After the laboratory tests of the units, they have divided the measurement campaign in two stages. One measurement campaign was executed near the factory,which is located at 300 mt on sea level in a valley with forests all around, so to have the equipment better under control and to proceed better to the optimization of the system. To check the data transmission quality of the system they deplaced it 10 km far from the factory at 650 mt on sea level.

The second measurement campaign was carried on real sorroundings and real working conditions.

They had to separate the flooding and the frost alarm units from the fire alarm and that , because they had to put the first by Pescara river, near the town of Pescara, and the fire alarm unit in the forest in Sulmona valley. They provided the flooding and frost units with their own photovoltaic current generator and transmitter. Because of the severe law restriction to set fire in the forest, they simulate the fires with lamp sets.

6. OPERATION AND MAINTENANCE

The flooding and frost alarm units transmit the acquired data to the Italian Hydrographic Institute of Pescara where the data are elaborated by the computer of the Institute.

The fire alarm unit transmits the data and images to the factory computer.

The forecast and early warning for frost worked in winter '84 with 98% certainty and in winter '84/'85 with 94% certainty. The unit worked ununterruptly.

Fig. VI gives an example of such forecasts. The forecast begins about 14 hours before the event. Normally it begins at about 2,00 p.m. for the coming night. The photovoltaic current generator has fed the system all the time. Some simulated critical cases have been performed by the staff by separating the modules from the batteries for 5 days. This operation has been repeated every 40 days, to simulate the total lack of sun. Because of the low temperature reached in February 1985 (- 15°C) the batteries charging regulator had some failures and the batteries liquid begun to freeze. It was necessary to insulate the container with polystyrene foam and some commercial electronic components had to be repleaced with military ones. The energy production of the photovoltaic modules dropped about 5% because of ice formation on the glass plate. The batteries efficiency suffered in this period very severe cold conditions, even if without noticeable variations in the performance.

The problems ecnountered for these low nocturnal temperatures disappeared during the morning because of the higher temperature, which was always above the 0°C.

With the fire alarm unit they worked, firstly, in short distance (100 mt), increasing,later, the distance to 10 km and to 15 km, to define the accuracy of the system, which has been stated with the following data: hot spots of 2,00 sq.mt at 15 km distance; 1,00 sq.mt. at 9,8 km and 0,50 sq.mt. at 5 km distance. During the months with snow they worked at the system to avoid false alarms, which can occur with low temperature due to sun reflection on the snow.

In Fig. VII they give the accuracy of the alarms. They have not defined a scala to distinguish the smoke stream of an hot spot in haze, fog or raining conditions yet. They are working on how to get it. They perform regularly check operations on the calibration of the sensors.

7. MONITORING

They have to monitor for the photovoltaic current generators the following parameters:

- solar radiation (Fig. VIII)
- voltage of the modules
- produced current of the modules
- energy production of the modules (Fig. IX)
- efficiency of the modules (Fig. X)
- voltage of the batteries
- current of the batteries
- voltage of the system
- absorbed current of the system
- absorbed energy of the system (Fig. XI)
- efficiency of the whole system (Fig. XII)

For the alarm system they control the following parameters:

- air temperature

- air humidity
- atmospheric pressure
- rain fall
- wind speed
- wind direction
- water level
- hot spot temperature
- smoke detection

For the solar radiation they use a pyranometer, which is calibrated by the makers with a champion pyranometer. The hydrometer has been calibrated with a mechanical one. All electrical magnitudes have been measured with a high accuracy multimeter.

All electronic parts have been tested and calibrated in a climatic chamber with static thermic tests at - 20°C and + 60°C for 60 hours in different humidity conditions, from 30 to 95%. Thermal cycle tests with temperature variation of 10°C each hour, for a total of 10 cycles. The wind with a frequencymeter.

They make the measurements every 15 min., the wind speed every 4 min., to get a reliable mean value, the mean wind speed and the maximal wind speed in the hour, being each of these measures associated with the wind direction.

Daily they calculate and print the min., max. and mean value of the day.

Every fortnight they elaborate the daily values to get the month values which are the basis of the annexed curves.

8. RELIABILITY OF THE SYSTEM

To understand better the reliability of the system they simulated with infrared lamps the different operation conditions on the field.

They operate the system trying to establish the dependance of the fire recognition reliability from four main parameters, which could influence negatively the results. These main parameters are, untill now:

a) the size of the fire (k_1)
b) the average diameter Ø of the log of the forest (k_2)
c) the average deepness d of the forest (k_3)
d) the average thikness T of the forest (k_4)

The distance at which the fire signal begins to have a sensible decreasing value is given by the following relationship:

$$D = \frac{l_g k_1 - (l_g k_2 + l_g k_3 + l_g k_4)}{k_1 - (k_2 + k_3 + k_4)}$$

The intensity of the signal at this distance will be:

$$I = I_o \left[\frac{k_1}{k_2+k_3+k_4} \right]^{\frac{1}{\frac{k_1}{k_2+k_3+k_4} - 1}}$$

The intensity loss is:

$$L(D) = I_o - \frac{I_o}{k_1 - (k_2+k_3+k_4)} \left[k_1 e^{-(k_2+k_3+k_4)D} - (k_2+k_3+k_4) e^{-k_1 D} \right]$$

9. EXPERIENCE OBTAINED DURING OPERATION

They shall speak only of the problems concerning the photovoltaic current generators.

It lacks, from the producers of the photovoltaic panels,for sufficient information concerning the efficiency of the panel in dependance of the panel temperature, so that they have a lot to do to establish the correct efficiency of the panels.

They usually have the characteristics printed on the technical leaflets and tehy assume that these ones are the results of long period measurements and that are useful in the dimensioning and in the design stage of the photovoltaic system.

It is obvious that to have the complete results,in dependance on daily excursion and monthly work,is helpful and of great interest for everyone working in photovoltaic application. For this reason it is important that the panel producers place more technical information and data about panel at disposal, so to be able to establish better basis of comparison with the results.

The same remark is adressed to the producers of batteries, but the problems met with the batteries are a little more complicate for the different parameters influencing the capacity and the efficiency of the batteries themselves. For instance, if they measure the energy input and output of a battery that won' t be sufficient to get enough information to state the efficiency of the battery itself. By that the necessity to have more data from the producers of the batteries.

Block scheme

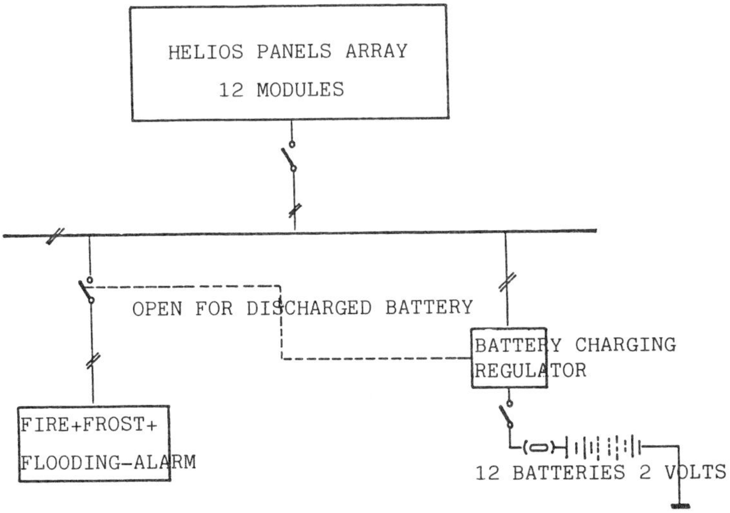

Figure 1

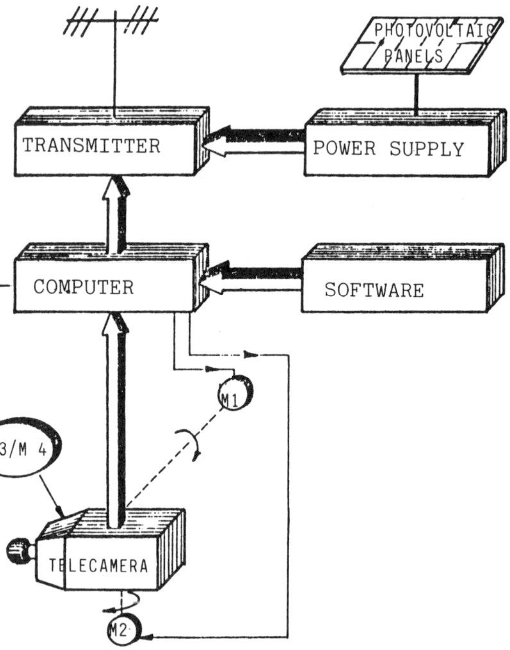

Figure 2 : Components of the fire early warning system

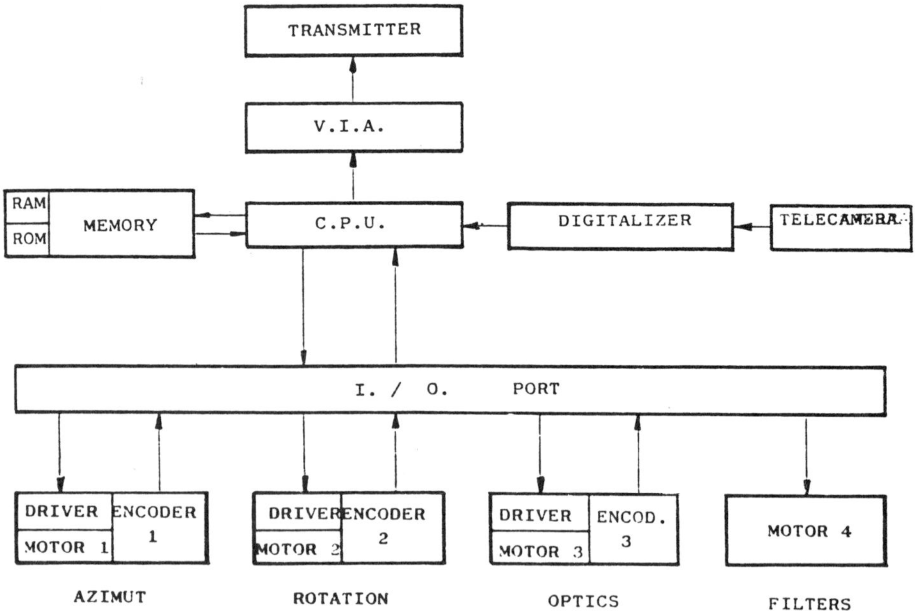

Figure 3 : Block diagram computer / Peripheral units

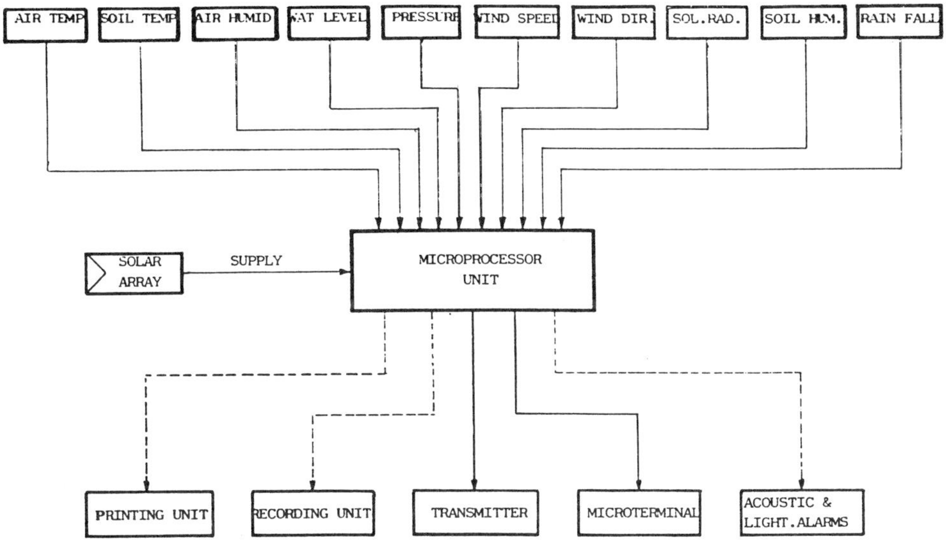

Figure 4 : Block scheme of the whole system

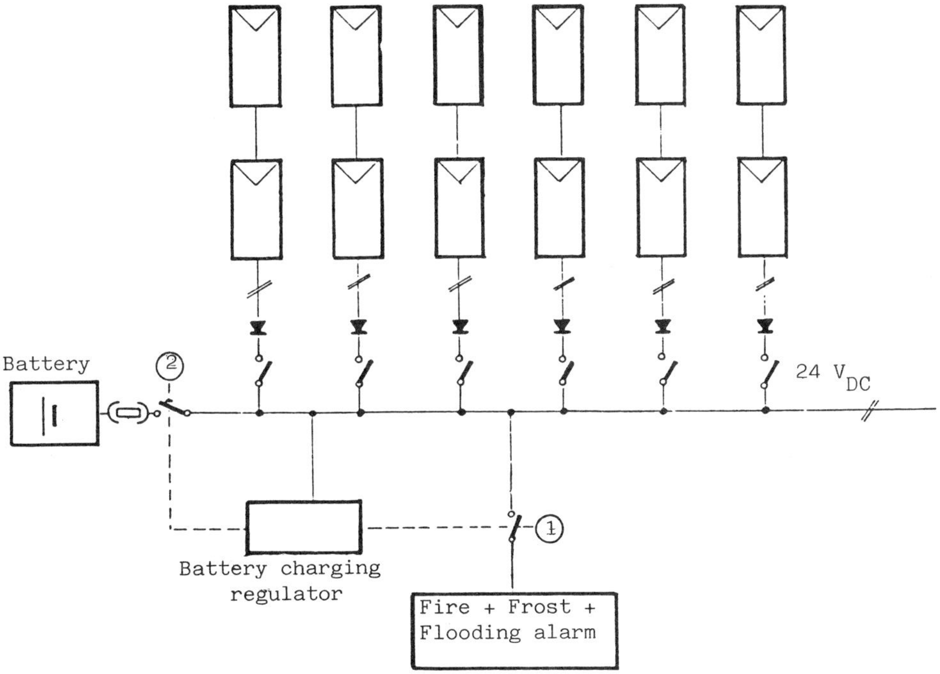

Figure 5 : Electrical scheme
1) Open for discharged battery
2) Open for overcharged battery

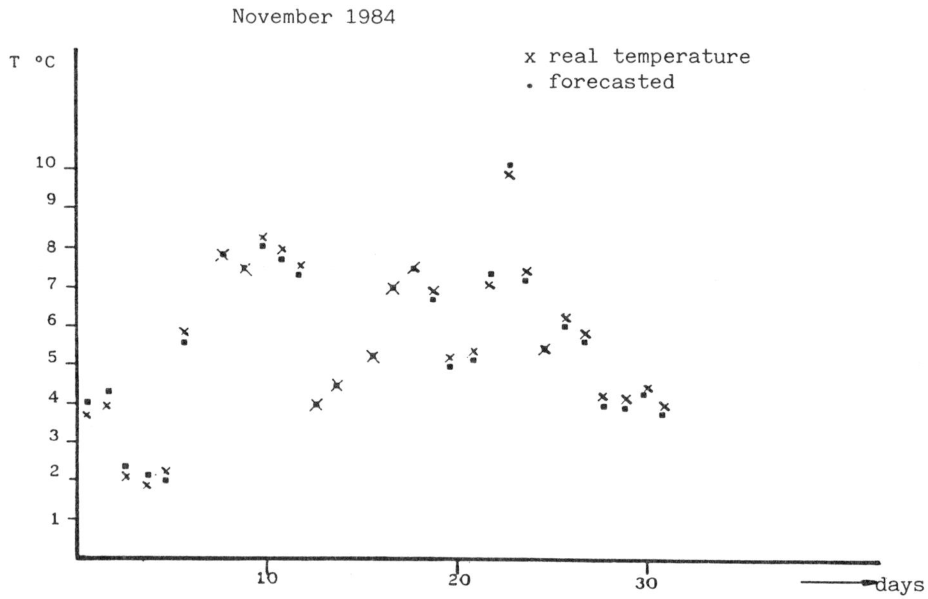

Figure 6 : Frost forecast

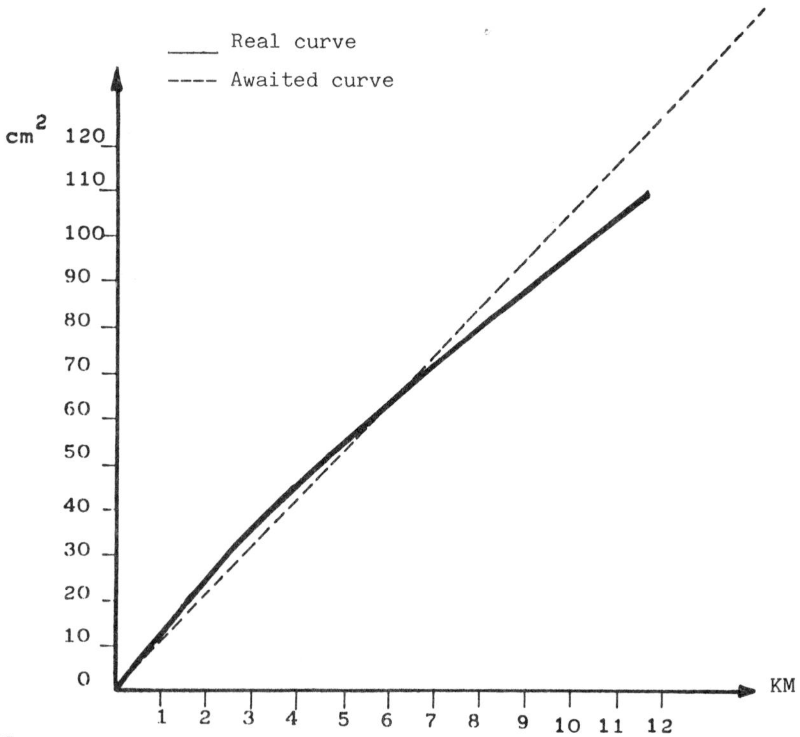

Figure 7

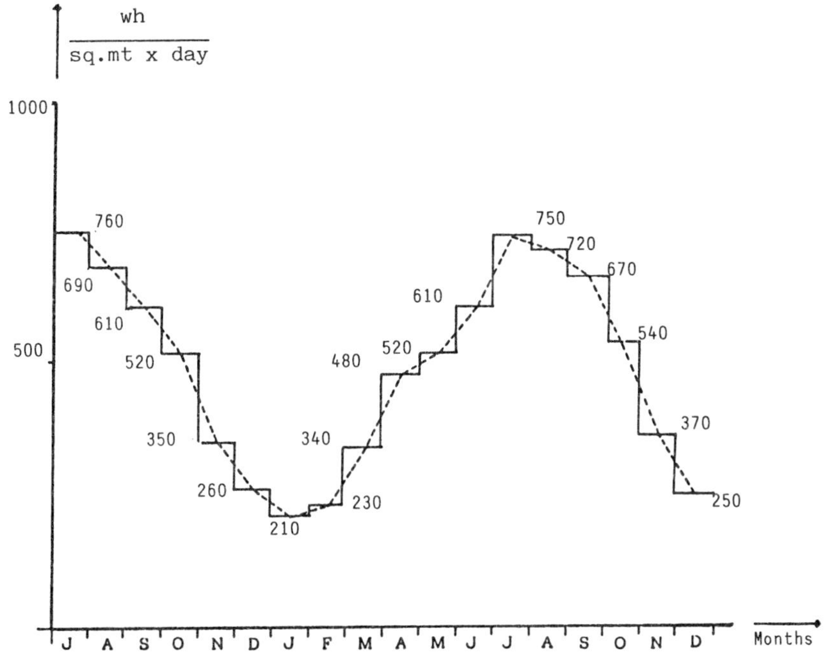

Figure 8 : Mean solar radiation in Wh/sq.mt x day
Period : July 1984 - December 1985

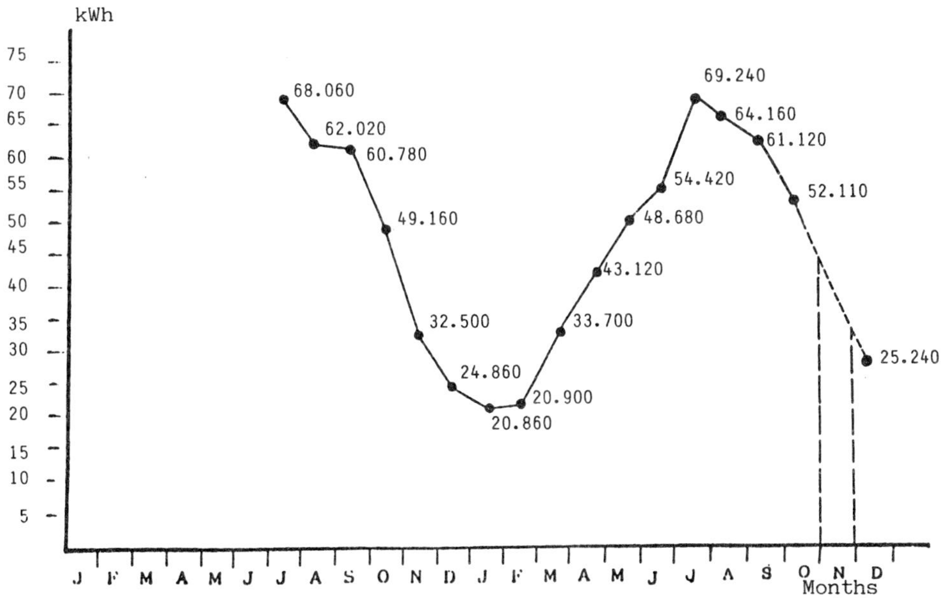

Figure 9 : Energy produced by the HELIOS H 436/1 Modules in kWh
Period : July 1984 - December 1985

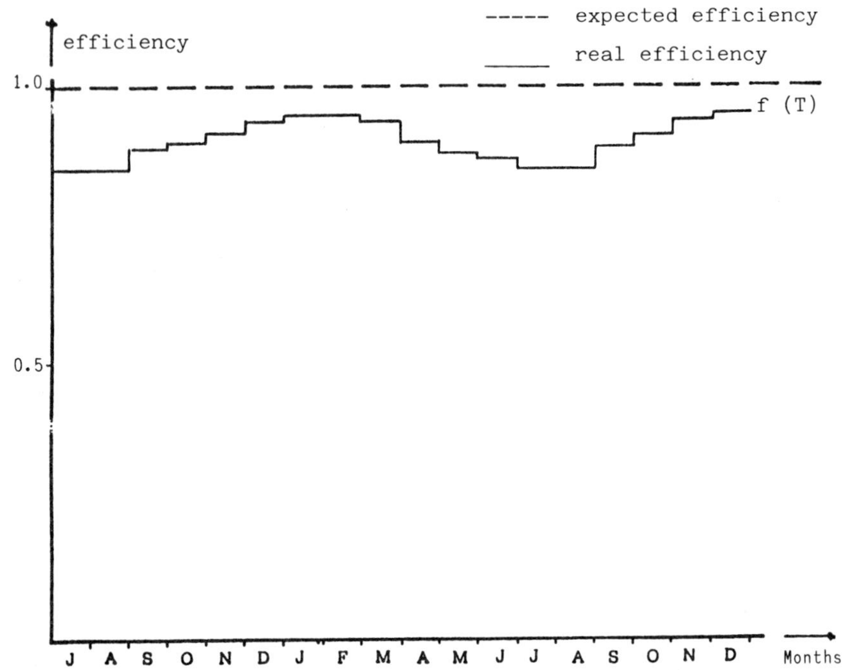

Figure 10 : Panel efficiency
Period : July 1984 - December 1985

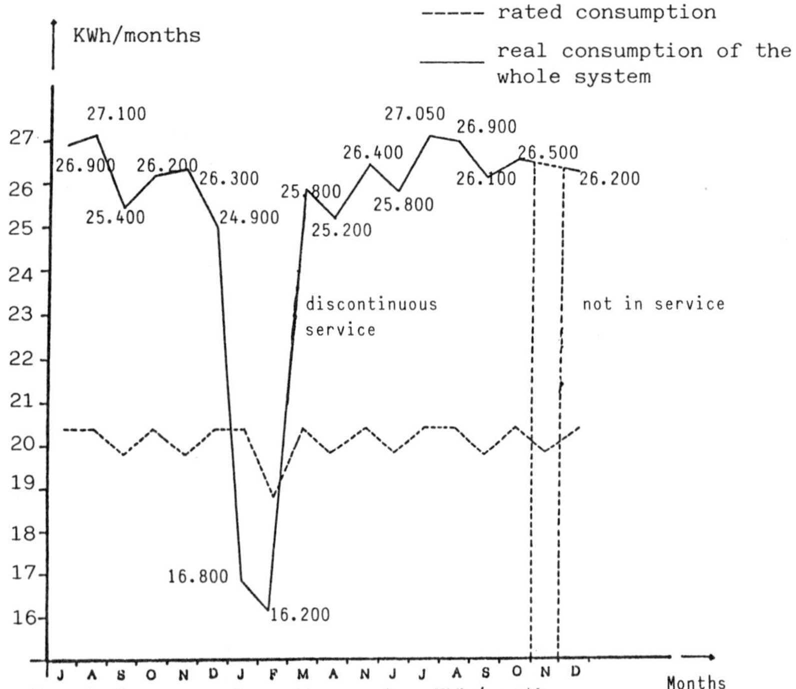

Figure 11 : Absorbed energy from the system KWh/month

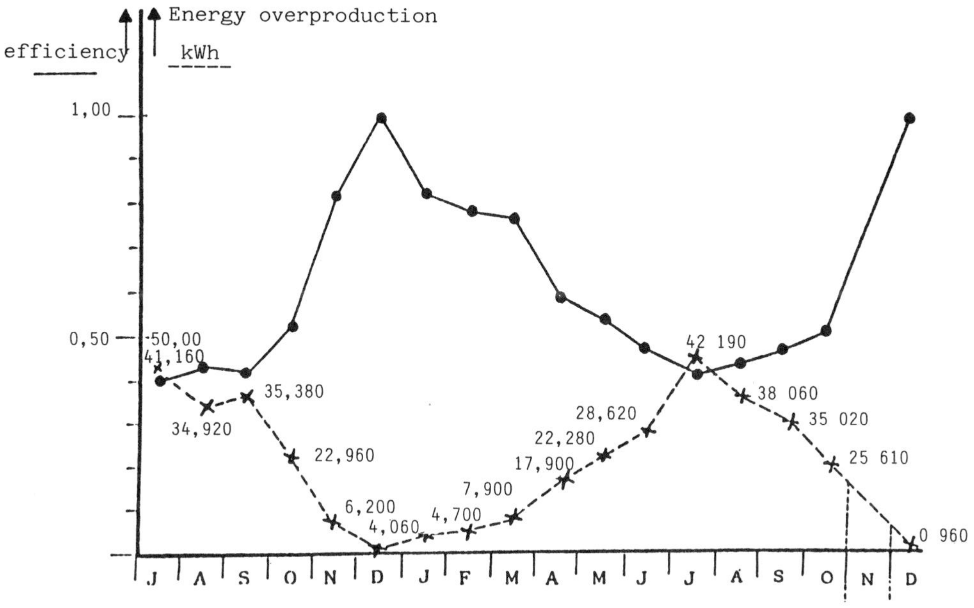

Figure 12 : Ratio between the energy required from the system and the energy delivered by the panel array
Period : July 1984 - December 1985

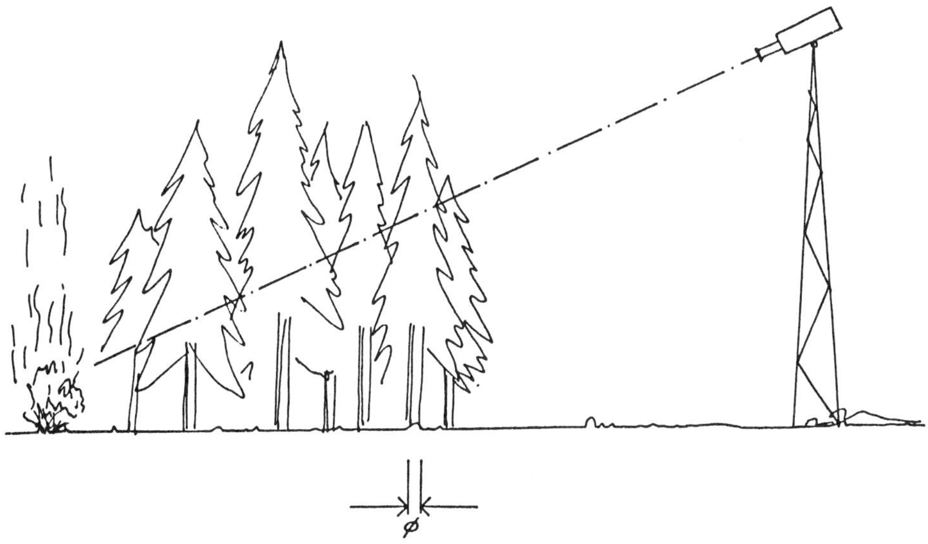

Figure 13 : Fire detection efficiency Vs. average diameters of the logs

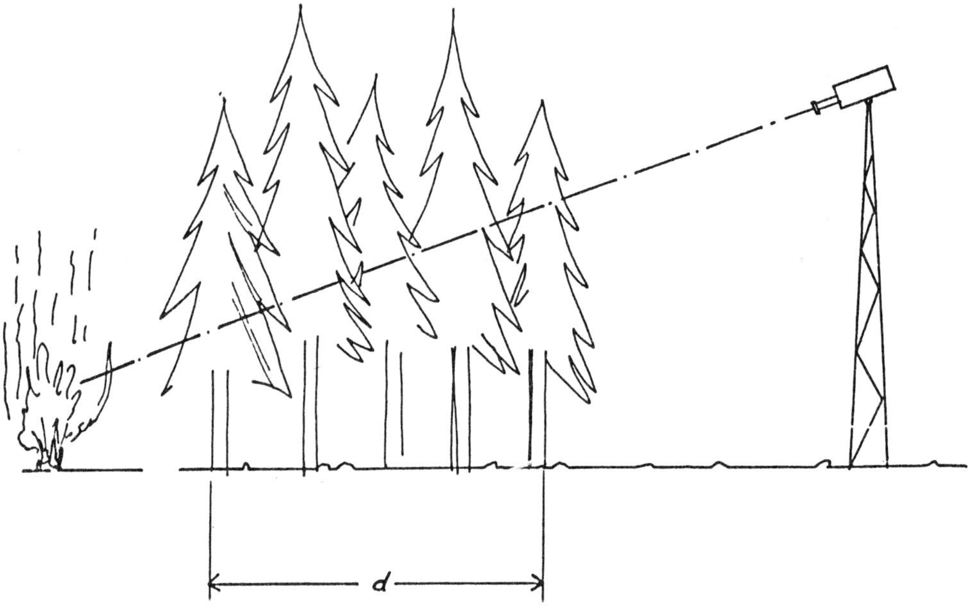

Figure 14 : Fire detection efficiency Vs. deepness of the forest

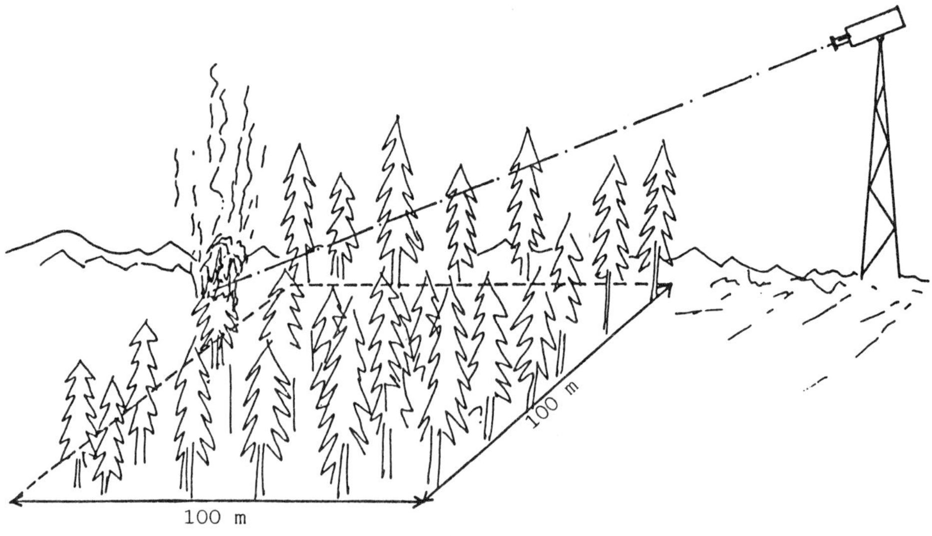

Figure 15 : Fire detection efficiency Vs. thickness of the forest (number of trees x 10,000 m^2)

Contract No. SE / 560 / 83

PHOTOVOLTAIC POWER SUPPLY FOR GROUND-WATER LEVEL MEASUREMENT

H.-J. Lowalt and E. Krausen
AEG Aktiengesellschaft, Anlagentechnik
Fachbereich Neue Technologien, Raumfahrt
Industriestrasse 29

D - 2000 Wedel / Holst.

SUMMARY

This project is worked out together with the HWW (Hamburger Wasserwerke, Hambourg Waterworks) to run water-level-meter-wells fully automatically. To determine the ground-water-level, nearby the pumping wells, is of great importance to guarantee a good drinking water quality.
The deterioration of many surface waters requires pumping of ground water, in this case from a depth of about 6 - 12 meters. The increased use of the ground water wells leads therefore to a more frequent inspection.

About 150 water-level meter wells are situated in the remote area of the marshland in the west of Hambourg.

In order to gather operation experience with such an automated measuring network, three level meter wells are equipped with photovoltaic power supply, to prove the reliability of such small stand alone photovoltaic systems for measuring purposes, not only for well meters, but for environmental measuring stations in general.

About 300 of these wells are equipped with mechanical control systems. A data collection is now done by a crew driving by car from well to well. This causes a considerable expenditure. By this means short-dated changes can not be realized. In the wintertime, at snowfall or other extrem weather conditions, a few level-meter-wells can not be reached at all - or can only be reached with increased effort. A data collection by a telephone line was rejected because of its high costs. So the decision was taken to install a radio system in the remote areas with a photovoltaic power supply.

1. INTRODUCTION

The HWW (Hamburger Wasserwerke, Hambourg Waterworks) are operating about 500 drinking water pumping wells for the supply fo the 1.6 million city and about 35 communities in the surrounding of Hambourg.

Sites of installation for these three pv powered data transmission systems were chosen, which are away from a grid supply, but very important for continuous data collection.

Because of protests by the landscape authority it was only possible to choose two sites in the marshland. To find a suitable location for the third equipment was a problem. So finally a site in Hambourg, about 30 kilometers from the receiver station was chosen.

The photovoltaic system consists of an approx. 310 Wp PV-array which is built up by 16 AEG - solar-modules of the type PQ 10/20, installed on one structure, a charge controller and an accumulator.
Data are collected by the central station via the transmitter station by all of these three stations. (see Fig. 1 and 2)

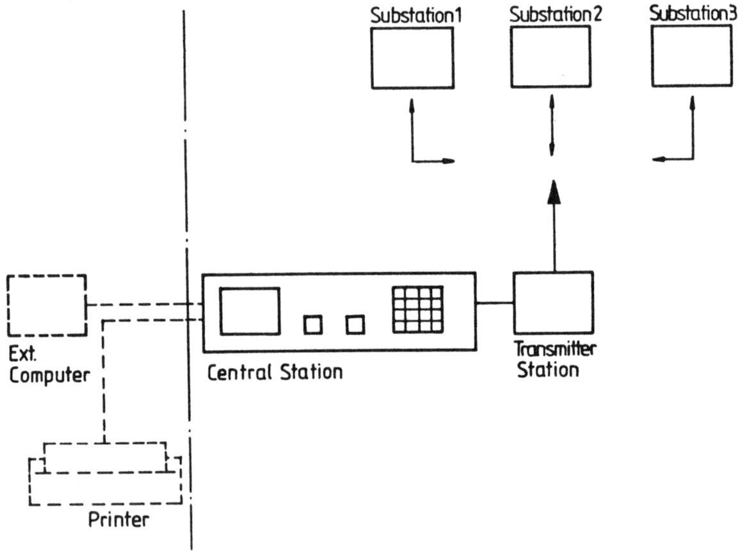

Figure 1

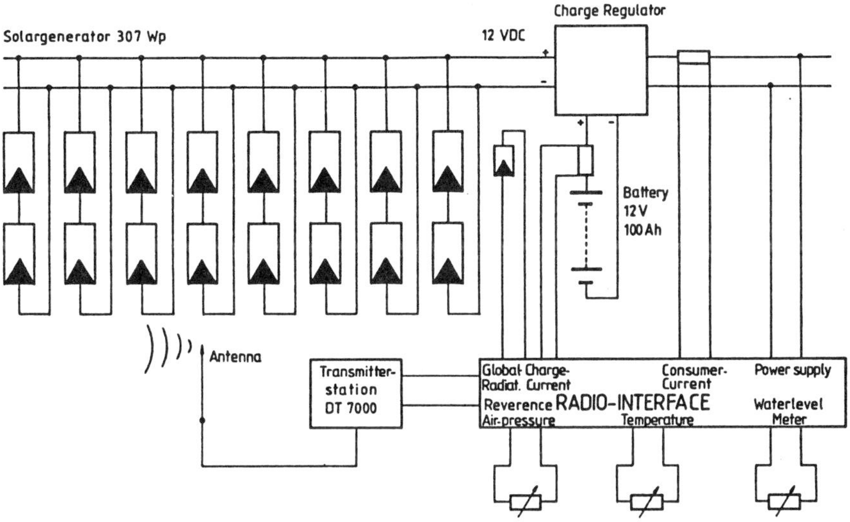

Figure 2

2. PHOTOVOLTAIC ARRAY

As main part of the array AEG modules type PQ 10/20 are used. This module consists of 20 multicrystalline silicon cells with the dimensions of 10 cm by 10 cm each. These cells are embedded between 2 glass sheets and plastic layers. The cell interconnection is done by a ultrasonic welding process. These "sandwich" is framed with stainless steel surroundings. The electrical and mechanical data are shown in the following data sheet.

Structural Data

PQ 10/20/01
PQ 10/20/02

Solar cells
Base material:
Multicrystalline silicon

Size
10 cm x 10 cm

Encapsulation
Glass/PVB/Glass
Stainless steel frame

Weight
3850 g

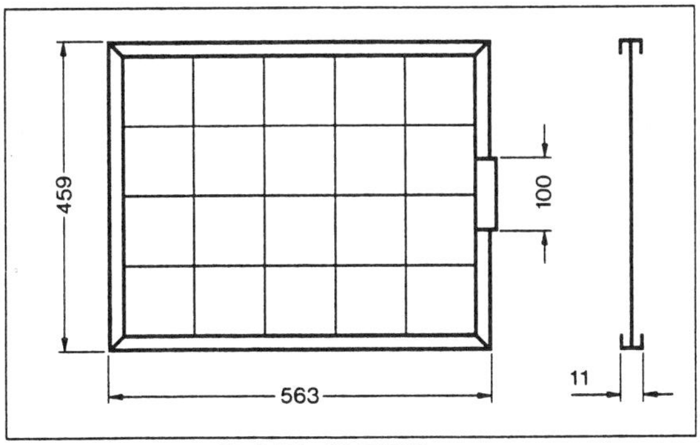

Maßbild/Dimensions (mm) PQ 10/20/01, PQ 10/20/02

Kenngrößen Characteristic values	(AM 1.5–100 mW/cm²)	Solarzellen-Temperatur Solar cell temperature 0 °C	 25 °C	 60 °C
Leerlaufspannung (V) Open circuit voltage (V)		12,3	11,2	9,6
Kurzschlußstrom (A) Short circuit current (A)		2,37	2,41	2,46
Strom im Punkt max. Leistung (A) Current at maximum Power (A)		2,18	2,20	2,23
Nennleistung (W) Nominal Power (W)		21,3	19,2	16,2
Max. Abweichung von der Nennleistung kleiner als 10% Max. difference to nominal power less than 10%				

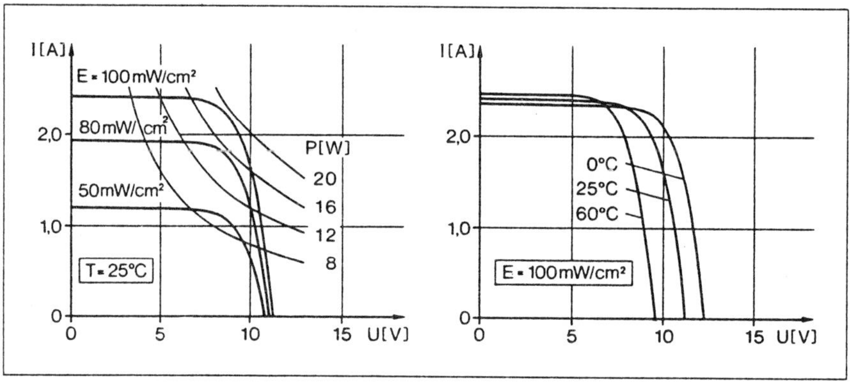

Temperaturabhängigkeit/Voltage, current and power data as a function of temperature				
Die Spannung/Voltage	erhöht sich/increases sinkt/decreases	um/by 0,4%/°C	unter/below über/above	25 °C
Der Strom/Current	erhöht sich/increases sinkt/decreases	um/by 0,06%/°C	unter/below über/above	25 °C
Die Leistung/Power	erhöht sich/increases sinkt/decreases	um/by 0,44%/°C	unter/below über/above	25 °C

The array is built up with standard structure components. The material for these structures is hot galvanized steel. Each station consists of 16 modules and all these 16 modules are installed on one structure unit. Due to the ground conditions in the marshland solid concrete foundations were necessary. The overall dimensions for one structure are about 2 meter length, 1.7 meter depth and about 2 meter height at a tilt angle of 50°.

3. CHARGE CONTROLLER

The approved charge regulators (type AEG BCR 12 Sh 300) for all three stations are of the shunt type with pulse-duration modulation. They regulate the battery after reaching the gassing voltage in order to prevent an overcharge and to reduce water losses of the electrolyte. Maintaining a preset gassing voltage limit leads to an automatic reduction of the charge current up to its float charging rate. The fact that lead acid batteries have a temperature characteristic of 6 mV/K/cell requires an adjustable gassing voltage. The devices offer both a fixed and a continuously adjustable overcharge voltage value. For a continuous tracking of the overcharge voltage an NTC-resistor is placed at the battery. For best performance it is fastened with an additional clamp directly on one of the battery terminals.

Operating a battery below the deep discharge voltage will influence its lifetime. Therefore a relay disconnects at this voltage level the consumer from the battery. The load is switched on again automatically after reaching a defined charge condition. To allow for a short-term voltage drop due to starting currents of motors or devices without disconnecting the load, the electronic has an adjustable time delay in the discharge protection circuit.

The battery type selected for this application is VARTA bloc. The capacity is 100 Ah at 12 V. This capacity represents a 14-day autonomous operation time.

Due to this very solid battery reserve capacity the afore-mentioned safety switching device will never be actuated during normal system operation. So a reliable power supply and system function is achieved.

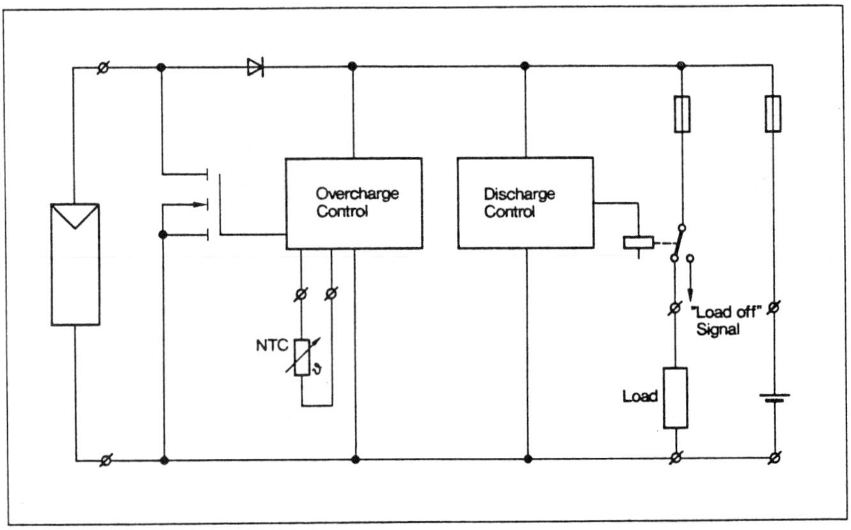

Basic Circuit Diagram

4. MEASURING AND DATA MONITORING SYSTEM

The system has been built up to collect and monitor data from the 3 measuring stations which are up to 30 kilometers away from the central station. All data will be transmitted by radio waves. In addition a voice contact can be realized to get in contact with the service man.

4.1 CENTRAL STATION

The main parts of the central station are a video data terminal, a radio installation with the amplifier and a stationary main control panel with the monitoring devices. Data of 32 stations in maximum can be collected. The measuring time scanning will be assigned and the individual meter stations will be automatically called upon the measurings. All transferred data will be checked upon transmission errors. In case of data transmission failures, the suitable substation will be requested for a new data transmission. The received data will be immediately turned into physical units and shown on a video display unit (VDU) so the water work operator is prepared to take the decision for switching over different pumping wells for equalization of water levels.

4.2 SUB STATIONS - DATA TRANSMISSION

These sub stations are installed directly at the level meter wells. They consist of the photovoltaic power supply, the radio interface with all sensors and measuring devices and the radio transmission station. The radio interface will be controlled by the computerized data system. It will be switched on by the radio contact of the central station, to fulfil the measurements and/or to transmit the collected data.

If there is a data transmission failure determined, the transmission will be repeated automatically by the request of the central station, as already mentioned, up to twice. (Ref. Fig. II)

4.3 MONITORING

In addition the central station will be equipped with a 5" video display unit. By means of video interface more video displays can be connected. One video screen page shows the data of one sub-station, containing 5 measuring values as well as date and time.

PRESENT STATUS

In consequence to the far distance of one project it turned out, that there was no reliable data transfer possible due to the nature of topography. We tried to remedy by variation of antenna height and position, but these measures resulted only in an insignificant improvement. As result we had to provide a stronger radio transmitter.

At all three stations the power supply, consisting of the photovoltaic array, the battery and charge controller, are installed and in operation. The two systems in the marshland are transmitting already data since October '86 to the central receiver station.

Together with the installation of the stronger radio equipment at the Hambourg site, in March '87, a few modifications had been done. The first practice with all three stations are satisfactorily and extremely helpfull for the operator. See following print output of actual data:

20.03.87 11:26:29		>>Akt.	Messwerte<<	
Stationsnummer:		1	2	3
Pegel	[m]	2.34	5.94	4.23
ILade	[A]	6.25	2.95	16.93
IVerb	[A]	0.220	0.805	0.222
UBatt	[V]	14.74	13.30	13.27
JGlob	[W/qm]	0.2	0.2	0.2
pLuft	[hPa]	1000.	1000.	1000.
Tint.	[Grad C]	3.28	4.67	2.59
Feuch	[%]	76.2	77.2	74.1
Text.	[Grad C]	4.94	3.87	4.23
URef.	[V]	2.50	2.50	2.50

Now only a few weeks of supervision and last adjustments are necessary to get the whole system including the data logger in reliable operation.

CONTRACT NO. SE / 267 / 84

PHOTOVOLTAIC POWER SUPPLY FOR THE BIRD AND WEATHER STATION ON SCHARHÖRN ISLAND

H.-J. Lowalt and E. Krausen
AEG Aktiengesellschaft, Anlagentechnik
Fachbereich Neue Technologien, Raumfahrt
Industriestrasse 29

D - 2000 Wedel / Holst.

SUMMARY

A very special region, caused by the exposed location, is the island of Scharhörn (0.15 km^2 of dunes) west of Neuwerk in the River Elbe Estuary. Located about 19 km off the coast across an expanse of low sandy tidal ground which is regularly flooded by the incoming tides it is not easily accessible. During frequent periods of inclement weather it is often not possible to bring supplies to the island. From March through November the island station is continuously occupied by one to three scientists. Because of the problems with the present diesel power supply, activities have been very limited. For instance up to now it has not been possible to install a deep-freezer to store collected birds for later studies relating to environmental toxicology. For the species protection programme there are other smaller electrical needs such as the illumination of insect traps. But all consumers can not be driven over a continuous time due to the noisy operation. So the decision was taken to install a Photovoltaic System. The system is built up with a 4.6 kWp solar generator, a battery of 100 Ah at 220 V and an inverter of 5 kVA.

1. INTRODUCTION

On this very special island, inside the Shoals National Park, a group of natural protection scientists (Jordsand Society) are stationed from March till November every year. Their main work is the observation of birds and their life studies relating to environmental contamination.

Additionally from time to time there are research projects being carried out on behalf of the Hambourg Board of Harbour and Works (Hamburgisches Hafen- und Bauamt).

The Hambourg Board of Harbour and Works operates on Scharhörn a weather station with a storm-tide warning service.
The Jordsand Society strives for research projects in cooperation with the Ornithological Station, Helgoland (Vogelwarte Helgoland) and the Hambourg Board of Natural Protection (Naturschutzamt Hamburg).

Due to limitations in operating the diesel power system they had to abandon a lot of purposes. So with the Photovoltaic System the lives of the scientists will be more comfortable on the island and a lot of more scientific investigation can be done.

The now installed p.v. system is designed to meet the average power requirements of all the research facilities during the months from March to September. This time span encompasses the critical months during which the adult birds mate, nest and nurture the young to independence. The diesel generator has been found to be extremely disruptive during this importang season.

To operate the whole range of industrial and commercial electrical apparatus from the solar generators it is necessary to convert the direct voltage generated by the solar generator into a sinusoidal alternating voltage by means of a static inverter (power converter). High availability is of particular importance for this application, therefore an industrial type of inverter (with slightly modifications) is used.

The AEG power converter for this solar system is a static inverter with transistors and built-in filter components. This system (see fig. I) converts the direct voltage generated by the solar generator into a 220 V, 50 Hz alternating voltage with the help of the AEG power converter. The excess energy is used to charge the storage battery. The charge regulator which is connected between the solar generator and the battery ensures an optimum charge of the storage battery. The storage battery contributes to the load current demand during the periods of time where the energy demand of the load exceeds the energy production capacity of the solar generator (e.g. night and during bad weather periods).

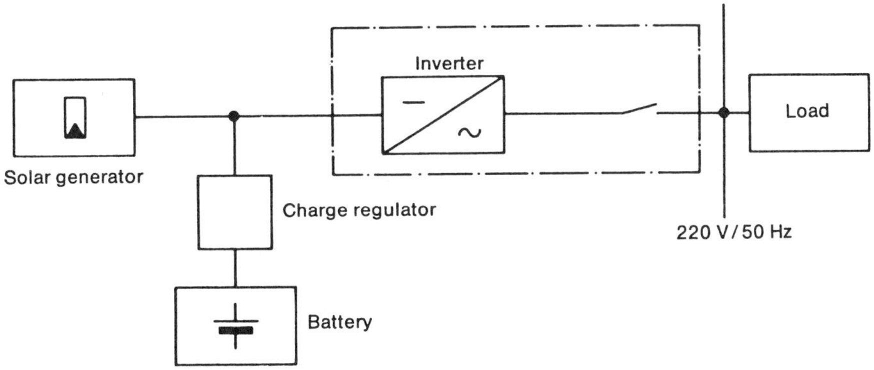

Figure 1: PV-Power Supply
Bloc Diagram

2. PHOTOVOLTAIC POWER SYSTEM

The PV array consists of 240 modules, type AEG PQ 10/20, which are mounted on a standardized hot galvanized steel structure. To form the string, for the nominal dc-voltage of 220 V, 30 modules are connected in series. 8 of these strings are connected in parallel resulting in a peak power of 4.6 kW.

The structures are mounted on a foundation which is built up with steel profiles bolted together. This foundation had been washed into the sandy ground and acting as a kind of anchor. Each of the afore-mentioned 8 strings of the PV array are connected individually via cables with the battery control system. This system controls the dc-voltage, and therefore the battery charging and discharging conditions by operational switching of these 8 strings. Also safety branches are built in for the protection of the battery and the inverter. A protection against a too high voltage will be done by a stepwise disconnection of all 8 strings. A protection against a deep discharge of the battery will be done by a stepwise disconnection of consumers according to their priority.

Fig. II: PV-ARRAY at the time of installation

The battery consists of 110 cells connected in series for the nominal DC voltage of 220 V. The type used is a OGi Bloc-Battery of Anker with a 10 h capacity of 100 Ah. As mentioned before as power converter a AEG transistor inverter is used with a nominal power of 5 kVA.

The inverter unit consists essentially of a bridge circuit with power transistors and appropriate circuit arrangements which make it possible to pulse the DC voltage at a higher frequency. In this way one achieves a sisusoidal output voltage without the expense of large filters.The whole power factor range, from cos phi = 0 capacitive through cos phi = 1 to cos phi = 0 inductive, is permissible.

This converter system is proof against short circuits. All the units necessary for the control, regulation and monitoring are contained in the control unit. The components are assembled on plug-in-cards and combined in 19 inch magazines.

The system is provided with a mimic diagram on the door. By means of LED's the operational state can be recognized quickly. In addition the control unit is equipped with a first failure alarm, i.e. the first failure which leads to the "FAULT" alarm, is additionally indicated by LED on the printed circuit board concerned. This makes easy fault detection possible.

The working range at the DC-side can be +/- 20 % of the nominal 220V. At every operation point inside this range the static AC-voltage variation is less than +/- 1 % over the whole load and power factor range. The dynamic AC-voltage variation for suddenly applied load 0 - 100 % - 0 is at least +/- 3 %. The efficiency at a cos phi of 0.8 is at full load not less than 93 %.

3. DATA MONITORING SYSTEM

The data monitoring system used is designed by AEG as a standard type according to the "Guidelines for Photovoltaic System Monitoring" by the Joint Research Centre, Ispra Establishment.

The universal data recording software has been built for the CMOS-microcomputer system CSS-SYS 800. Thereby the recording of measured values via 16 analog input channels with a resolution of 12 bit and additional signs is possible. Here the raw data of these 16 input channels are converted according to the regulation defined by the user in the system configuration, are connected and assigned to a display-respectively recording channel.

3.1 TECHNICAL DATA

Measuring Channel

- 16 channels with common reference point or
  8 channels differential
- resolution 12 bit and signs
- measuring voltage range +/- mV up to +/- 11 V adjustable
- input card with precision resistances for measurement purposes for use of measuring transformer with current output 0 - 10 mA of +/- 20 mA.

Measuring Times

- reading cycle for mean value formation programmable 5 - 120 sec
- recording cycle for mean value data programmable up to 3600 sec (lower limit of the cycle dependent upon the sum formation cycle)
- recording duration at least 80 days with max. usage of 20 recording channels (diskette)

Electrial / Mechanical Construction

- 19" rack - 3 height units
- modular construction with Europe cards
- all cards are set up largely in CMOS technology
- vacant plug-in places for extensions

Operator Interface

- standard video terminal as WYSE 30 or TELEVIDEO 925 with RS 232 interface
- hand terminal with LCD - display

CONTRACT NR: SE 692 - 84

PHOTOVOLTAIC LIGHTING OF ARCHEOLOGICAL SITES

F. Basevi, E. Bricca, M. Orsoni
PRAGMA S.p.A.
Via A. D'Andrea, 6 I - 00048 Nettuno

SUMMARY

Lighting and auxiliary services powering for archeological sites in Tuscany will be provided by means of photovoltaic systems. A 19,800 Wp PV array will power the indoor and outdoor lighting system of Cetona caves, few kilometres from Cetona village, while a 6,600 Wp PV array will feed the external lighting of Ildebranda Tomb near Sovana village and, when needed, the sump pump for rain drainage of the underground chambers.

Presently detailed design is completed, as well as purchasing procedures, while installation activities are still to start owing to delays in obtaining permits from Environmental Protection and Fine Arts authorities.

1. INTRODUCTION

Objective of this project was to illuminate two extremely interesting archeological sites, in the mounty part of the Etruscian area in Tuscany, not connected to the national utility grid, in order to allow turists to visit them and to prevent one of them from being regularly floded by rainfalls.

Cetona caves openings, on a steep hillside, lead to several grottos in sequence where human findings of previllanovian period were encountered and an Etruscian necropolys. Internal lighting will be provided in four grottos during the day, while from twilight on, the path leading to the caves and caves openings will be illuminated for a few hours.

The second site, Ildebranda tomb, belongs to an important Etruscian necropolys near Sorano and is of high aristic and archeological interest. The front of the tomb, carved in the hillrock, will be lighted from twilight on, while the mortuary chamber will be illuminated during the day. When needed, a sump pump will operate to extract rain water from the mortuary chamber itself that is situated below ground level. The photovoltaic systems are modular, the Cetona one being exactly three times the Ildebranda one, and consist of the following main components:

	Cetona	Sovano
photovoltaic array (Wp)	19,800	6,600
storage batteries (Ah)	700	300
DC/AC converter (VA)	5,000	3,500
control board	1	1
lights (nr)	37	17
water pump (nr)	-	1

2. PHOTOVOLTAIC ARRAY

The photovoltaic arrays consists of 72 DL 2x36 PRAGMA modules, connected in strings of 25 modules each, with the following characteristics:

	Cetona	Sovano
modules (nr)	300	100
strings (nr)	12	4
subfields (nr)	3	1
open circuit voltage (V)	525	525
short circuit current (A)	54	18
nominal voltage (V)	340	340
tilt angle (deg)	35	35

3. POWER CONDITIONING

The storage was designed to guarantee 5 days of autonomous operation in "no sun" conditions for both installations. In Cetona it consists of 150 lead acid elements FIAMM SMBF/7 type, series connected to obtain a nominal voltage of 300V and a capacity of 700 Ah (over 10 hrs). In Sovano 150 elements FIAMM SMBF/6 type connected in the same way, provide on overall capacity of 300 Ah (over 10 hrs).

Three charge regulators handle the power from the twelve array strings in Cetona while one is sufficient for the Sovano array. In the control cabinet switches and overvoltage and surge protections are also housed.

The DC/AC static converters, similar in design but not in size, have the following features:

	Cetona	Sovano
rated output (kVA, cosØ0,8)	5	3,5
maximum output (kVA)	7,5	5,2
input voltage	300 V, DC, + 20%, - 10%	
output voltage	220 V, ± 2%	
frequency	50 Hz, ± 0,5 Hz	
efficiency at nominal load	> 75%	

4. DETAILS OF PV USER APPLIANCES

Twentyfour incandescent 40 W lamps in waterproof casing will be installed inside the Cetona caves, while thirteen 40W fluorescent ones will be installed outside; incandescent lights, although with lower louminous efficiency, have been chosen because of better quality of light and their better resistance to moist environment.

In Sorano two 250W spot lamps will floodlight the tomb facade while six 40W incandescent lights will illuminate the underground rooms; nine flourescent 40W lamps will light up the surroundings and the auxiliaries room. A 250W sump pump 220V, 5oHz is also provided,to drain rain water when necessary.

5. MONITORING SCHEME

At Cetona the following parametra will be monitored and recorded:

	Units
Insolation, global (horizontal)	kWh/m^2
Insolation, array plane	kWh/m^2

Max. wind speed at 10m height	m/s
Wind direction at 10m height	deg
Max ambient temperature	degC
Max module temperature	degC
Array output energy	kWh
Battery charging or discharging current	A
Energy to and from batteries	kWh
Inverter DC energy	kWh
Inverter AC energy	kWh
Energy to DC loads	kWh
Energy to AC loads	kWh

as required by the "Guidelines for Photovoltaic System Monitoring" issued by Joint Research Centre and Directorate General for Energy of CEC, January 1985.

In addition to that the following data will also be recorded:

	Units
Circuit opening for max voltage array/battery	YES/NO
Circuit opening for max voltage battery/loads	YES/NO
Battery voltage	V
Array current	A
Current to the load	A
Current to the 5 kVA inverter	A

6. PRESENT STATUS

Presently the detailed design and procurement phases have been completed. Unfortunately obtaing installation permits from regional Environment Protection and Arts and Archeological Sites Administrations took much longer than expected. Foundations had to be redesigned in order to avoid altering the ground in anyway and now are simply settled on it.

Installation activities are expected to start and be completed during summer 1987.

SESSION IV - GENERAL AND CONCLUSIONS

Overview of Session IV

Monitoring results from demonstration projects

European market for local electricity production for isolated areas

DC appliances for use with PV generators

Points arising during discussions

Conclusions

SESSION 4 : GENERAL AND CONCLUSIONS

4.1 OVERVIEW OF SESSION 4

4.1.1 Monitoring Results from Demonstration Projects

Mr Blaesser presented an analysis of the monitoring results received by the JRC, Ispra from the demonstration project in Antikythira (SE/704/83) and from the project involving the electrification of mountain refuges in the French Alps (SE/696/84).

The Antikythira PV powered relay station has been analytically monitored, and the monitoring results indicate that PV is a reliable power source for such remote telecommunication relay stations. The mountain refuges in the French Alps have been globally monitored, ie only integrated values of the energy produced by the array and energy consumed by the user have been recorded. The results indicate that the daily energy consumption in each refuge for each month is relatively consistent.

The collection and analysis of monitoring results was shown to provide a valuable basis for demonstrating the economic viability of PV systems, and to provide useful information on system operation which can be used to assist with the design of future systems.

4.1.2 PV Electrification of isolated sites

Mr Drouot from Europe Energies presented the results of a study on the electrification of isolated sites. Approximately 1 million people in Europe live in unelectrified dwellings. 80% of these live in the southern Mediterranean countries, especially Italy, Spain and Portugal. About 95% of this group of 1 million would like to be electrified, and most would prefer to be grid connected.

The study included a review of the legislation and finance covering rural electrification. This was found to be inadequate except in Italy where rural electrification receives a 100% subsidy.

The costs of grid connection and PV generators were compared, and it was found that PV became competitive when the site was a few km from the grid. The main alternatives to PV and grid connection are diesel and wind generators. Diesel generators are currently very popular as there is a good marketing network, and secondhand generators are easy to buy and sell.

It was concluded that the total potential for PV electrification in Europe was between 19 and 133 MWp. However, Mr Drouot believed that this market will only be realised if 100% subsidies are available for PV systems, as is traditionally the case for grid connections in rural areas.

4.1.3 DC Appliances for use with PV generators

The results of a brief survey of manufacturers of dc appliances were presented by Dr Gillett from Sir William Halcrow and Partners. Companies in Europe, America and Canada, from the PV industry, leisure industry and electrical equipment industry had been contacted and asked for information on the low voltage (12V or 24V) dc appliances which they supplied. It was found that high efficiency, lighting with a low power consumption was available fairly easily, as were small low power TV sets. DC refrigerators were more difficult to obtain and varied widely in efficiency and price. Various other small dc appliances were also available.

In general, dc appliances are more difficult to obtain than equivalent ac appliances and the choice for consumers is more limited. DC appliances are mostly available from small companies, and their products are aimed mainly at the leisure market. Many of these companies were unable to supply information on the energy consumption of the appliances. It was concluded that the labelling of products with their power rating should be encouraged so that consumers can make an informed choice.

MONITORING RESULTS FROM DEMONSTRATION PROJECTS

by G.Blaesser

Commission of the European Communities
Joint Research Centre - Ispra Establishment - Esti
21020 ISPRA (VA) Italy

SUMMARY: This report gives an overview of the monitoring data obtained so far from the demonstration projects. It discusses in detail the results from the Antikythira plant and from a group of mountain refuges in the French Alps.

INTRODUCTION

Since the start of the PV demonstration program sponsored by DG XVII we have received the following monitoring data:

- Projects with analytical monitoring:
    - Antikythira: from Oct.1985 to Nov.1986
    - Karpatos: Nov.1986
    - Freiburg: Sept. and Oct.1986
- Projects with global monitoring:
    - Rural electrification: 9 houses, June 1985 to Apr.1986
    - Mountain refuges: data from 6 sites, June 1986 to Oct. 1986

Since the period covered by the Karpatos and the Freiburg projects is not yet long enough to judge the performance of these installations (though the data available look rather promising) I shall discuss here only the results from the Antikythira project and those of the mountain refuges in the French Alps. The very interesting results from the rural electrification (40 houses in southern France) have already been analysed thoroughly by the AFME (1).

1. ANTIKYTHIRA

1.1. Introduction:

The Antikythira project provides PV energy to a communication link between the isle of Crete and the Greek mainland. The relay station is installed on the island Antikythira (35.85 N, 23.30 E). The station is owned by the Hellenic Telecommunication Organization (OTE).

The PV generator reduces the fuel requirements of the Diesel generators, which had powered the relay station up to now. Since the transport of the fuel presents logistical problems during a large part of the year which increase the running costs this substitution seems to be economically attractive.

1.2. Description of the installation:

The PV field has a nominal peak power Pp = 20.8 kW and consists of 8 subfields of 2.6 kW (each of 5*4 = 20 modules of type SM 144-09 with a Pp = 130 W and a surface of 102*42 cm*cm). 8 batteries with a capacity of 1700 Ah each provide storage for ca 8-10 days. 8 DC/DC converters with maximum power point trackers adapt the field to the load. The Diesel generators remain available for back-up.

Monitoring is provided by a HP9915 computer via an HP3421 Data Acquisition and Control Unit. The data are written by a tape-drive CDS2.50 unto ECMA-46 cassettes and sent to ESTI.

1.3. Summary of the data:

The results of one year of monitoring are summarized in Table I. Here, mon-time gives the total number of hours for which monitoring data were recorded, EAt is the integrated energy output from the PV array field and PVU is the energy from the PV-field effectively used in the load (transmitter). The Capacity Factor is

Cap.F. = PVU/(Pp*mon-time) = (mean power to load)/(peak power)

i.e. a capacity factor of 10% corresponds in this case (Pp = 21 kW) to a mean load power of 2.1 kW and represents a fairly standard value for solar powered telecommunication installations.

Mean array efficiency is

Mean Eff. = EAt(month)/((monthly integrated irradiance)*Ftot)

while maximum array efficiency is the maximum of such a ratio computed for each hour. The Arr.Eff.(max) data are typical for the type of modules used and were found also in other PV installations (and in the EARS field of ESTI) for the same modules. The mean array efficiency may be considerable lower than the maximum array efficiency in cases of bad load management. The present case is an example of excellent operation procedures, as the mean efficiencies are only ca 10% less than the maximum efficiencies. During the sommer months the efficiencies are lower than in winter time because of the temperature effect.

The tilt efficiency is the ratio of integrated irradiance in the array plane to horizontal integrated irradiance: since the inclination of the modules is chosen in such a way as to absorb more energy during the winter months, the tilt efficieny is higher than one in winter and lower than one in summer.

The converter efficiency is the ratio of energy output to energy input at the DC/DC converter which transforms the operating voltage of the PV-field to that used by the transmitter.

The overall efficiency is the ratio of PVU to the total integrated horizontal solar energy, i.e. it is

PVU/((global horizontal monthly irradiation)*Ftot)

It also clearly reflects the seasonal trend of the tilt efficiency.

1.4. Conclusion:

Apart from some initial problems during the set-up phase and problems with the irradiance measurements during January 1985 the monitoring results from the Antikythira station are of good quality and show that solar energy is a reliable power source for remote telecommunication relays.

2. ALPINE REFUGES

2.1. Introduction:

A number of PV generators have been installed in mountain refuges in the French Alps. They are situated between Geneva and Nice and operated by the society GTA ("Grande Traversee des Alpes"). Six of these projects have provided monitoring data covering part of the summer season of 1986.

2.2. PV Generator and Monitoring System:

Three types (A,B and C) of installations are foreseen: those of type A correspond to houses with ca. 30 places, type B to ca. 50 places and type C to ca. 80 places.

All installations used polycristalline silicon modules PW 402 from Photowatt, with a nominal peak power of 40 Watts. In several cases two subarrays were installed with different orientation. The total number of modules installed is 7 for sites of type A, 12 for those of type B and 15 for type C sites.

The installed number of batteries (12 V) for the three types is 7 (150 Ah each), 12 (125 Ah) and 15 (125 Ah), which corresponds to an energy capacity of 12.6 kWh, 18 kWh and 22.5 kWh.

For monitoring and display of the integrated energy production and consumption Wh-counters(DC) of type AMPERE-TEC 1040 (SCHLUMBERGER - ENERTEC) were used . Normally, the guardian recorded these readings every evening on a data sheet, but sometimes the data are available only for much longer intervals.

While the total energy consumption was monitored directly, production was measured only at one module (or one module per subarray in cases where 2 subarrays with different orientation were used), and then multiplied by the number of modules per subarray. If the module differs in its characteristics from the mean of the array, this procedure may lead to a systematic over- or underevaluation of the production; normally such an error should be less than 10%.

2.3. Evaluation Procedure:

The reported data were sorted by months. In the case of Nice, only the sums were reported for a period from Aug.12 to Sept.25; here, the values were distributed between August and September in a ratio 4:5, but these data were not used for the determination of statitical parameters such as standard errors. In other cases, where sometimes sums over two days were given, such sums were divided equally for both days and the resulting data included in the statistical evaluation.

The quantities given in the table are defined as follows: Days of monitoring are the total number of days per month over which the energy consumption was integrated. Thus the mean daily consumption is obtained by dividing the consumption as given by the energy counter by this number of days. When sufficient daily data were available, the standard error of <X> was computed as

(1) std.err. = sqrt(<(X-<X>)*(X-<X>)>/(N-1))

where N is the number of days used in averaging.

The capacity factor - which is the ratio of mean power to nominal (peak) power - was calculated from the mean daily consumption by dividing it by the installed nominal power (40 W times total number of modules in both subarrays) and by 24 hours/day.

In some cases the Wh-counters for the production were installed later than the counters for the consumption. In these cases the mean production was calculated by dividing the indicated integral production by the number of days N during which that counter was operational. This number was also used for the computation of the standard error.

The ratio consumption/production was calculated by dividing the mean daily consumption by the sum of the mean daily productions of both subarrays.

2.4. Results:

Even though no irradiation data have been reported it is obvious that the climatic situation of Bossetan and Emparis is not so favorable as that of Nice, Soreiller and La Pare. In the case of the last three stations the energy production exceeds the consumption almost by a factor of two, whereas for the other group the available amount of sunshine can hardly satisfy the demand. The low values of the capacity factor (of the order 2-3%) at high consumption/production ratios (of the order of one, even taking into account all uncertainties) clearly indicate this fact.

The "sunshine" stations reach capacity factors of 5% and higher for consumption/production ratios of 0.5. Either by reducing the size of the array or by increasing the consumption one could easily operate these stations at a capacity of 10%.

Anterne is in an intermediate solution. It is using all its energy but reaches a capacity factors above 4%. Thus even in a less favorable climate it gets as much useful energy per installed Wp as stations such as Soreiller or La Pare.

2.5. Conclusions:

From this analysis it can be seen that even rather "global" monitoring by using only few Wh-counters can lead to results which will be important for a better analysis of the economics of PV systems. In particular, it seems that the amount of energy use is fairly predictable (as may be seen also from the fact that the relative standard error of this quantity is only ca 10% when the data are sampled over a whole month) so that with sufficient knowledge of the meteorological conditions at the site the design of the installation may be well adapted to the needs.

REFERENCES

(1) CAMPAGNE DE MESURES SUR L'OPERATION "40 MAISONS PHOTOVOLTAIQUES" by E.FUGIER ("Rapport de stage" at AFME-Valbonne) IUT Genie Thermique et Energie - GRENOBLE - June 1986

TABLE I: SUMMARY OF ANTIKYTHIRA MONITORING

Month:	nov.85	dec.85	jan.86	mar.86	apr.86	jul.86	aug.86	sep.86	oct.86
mon-time(h)	687	537	337	173	46	175	334	648	397
EAt (kWh)	1578	1137	321	172	120	528	1054	1821	1124
PV_U (kWh)	1380	1016	278	140	90	372	713	916	853
Diesel(kWh)	0	3	433	0	0	0	3	0	0
Capacity F.%	9.6	9.0	(3.9)	9.1	9.3	10.1	10.2	6.7	10.2
A.Eff.(max)%	7.2	7.1	(2.8)	6.1	5.8	6.0	6.2	7.0	6.8
A.Ef.(mean)%	6.6	6.4	(1.7)	5.4	5.3	5.5	5.1	5.5	5.3
Tilt Eff.%	163	(352)	(600)	109	94	79	95	111	128
Convtr Eff.%	78	78	(113)	87	90	91	92	91	92
Ov.all Eff.%	9.4	(20.2)	8.9	4.8	3.7	3.0	3.3	3.1	5.1

TABLE II: SUMMARY OF PV MONITORING RESULTS FOR ALPINE REFUGES (SUMMER 1986)

Bossetan (80 places): 9 modules facing S, 6 modules facing E;

month:	June	July	Aug.	Sept.	Oct.
days of monitoring	7	31	31	30	3
consumption (kWh)	2.82	12.74	13.13	6.34	0.11
mean daily c. (Wh)	403	411	423	211	37
std.err(%)	14	10	11	12	86
Capacity Factor(%)	2.8	2.9	2.9	1.5	
Production S (kWh)	1.73	5.74	6.16	3.36	0.37
Production E (kWh)	1.55	5.07	5.51	1.64	0.45
mean daily p.S (Wh)	247	185	199	112	123
std.err(%)	28	13	22	15	12
mean daily p.E (Wh)	221	163	178	55	15
std.err(%)	24	12	13	18	90
consump/product. (%)	86	118	113	127	27

Nice (90 places): 6 modules facing S, 9 modules facing SE;

month:	June	July	Aug.	Sept.	Oct.
days of monitoring	20	31	31	30	31
consumption (kWh)	21.23	23.18	28.60	22.67	12.19
mean daily c. (Wh)	1061	748	923	755	393
std.err(%)	9	7			11
Capacity Factor(%)	7.3	5.2	6.4	5.2	2.7
Production S (kWh)		7.92	27.10	22.55	9.38
Production SE(kWh)		4.71	26.01	20.15	12.56
mean daily p.S (Wh)		880	874	752	303
std.err(%)		29			24
mean daily p.SE(Wh)		523	839	672	405
std.err(%)		43			8
consump/product. (%)		53	54	53	56

Soreiller(110 places): 15 modules facing S;

month:	Aug.	Sept.
days of monitoring	30	13
consumption (kWh)	18.11	5.92
mean daily c. (Wh)	604	455
std.err(%)	11	
Capacity Factor(%)	4.2	5.7
Production (kWh)		9.62
mean daily p. (Wh)		740
std.err(%)		
consump/product. (%)		62

TABLE II (cont):

La Pare (28 places): 7 modules facing S;

month:	Aug.	Sept.
days of monitoring	6	14
consumption (kWh)	1.27	4.59
mean daily c. (Wh)	212	328
std.err(%)	22	15
Capacity Factor(%)	3.2	4.9
Production S (kWh)	3.69	9.69
mean daily p.S (Wh)	615	692
std.err(%)	26	9
consump/product. (%)	34	47

Anterne(47 places): 7 modules facing S;

month:	July	Aug.
days of monitoring	27	31
consumption (kWh)	8.19	7.40
mean daily c. (Wh)	303	239
std.err(%)	19	10
Capacity Factor(%)	4.5	3.6
Production S (kWh)	7.29	8.39
mean daily p.S (Wh)	270	270
std.err(%)	11	8
consump/product. (%)	112	88

Emparis(40 places): 9 modules facing SE, 3 modules facing SW;

month:	Aug.	Sept.
days of monitoring	30	30
consumption (kWh)	7.55	5.70
mean daily c. (Wh)	252	190
std.err(%)	4	10
Capacity Factor(%)	2.2	1.6
Production SE (kWh)	6.07	6.71
mean daily p.SE(Wh)	202	224
std.err(%)	8	8
Production SW (kWh)	0.26	2.86
mean daily p.SW(kWh)	9	95
std.err(%)	(100)	12
consump/product. (%)	119	60

Contract n°: 85-B-7032-11-002-17

EUROPEAN MARKET FOR LOCAL ELECTRICITY PRODUCTION FOR ISOLATED AREAS

L P Drouot and J-G P Napoléoni
Europe-Energie Conseil
4, rue René Barthélémy
F-92120 Montrouge, France

SUMMARY

The total European population that live permanently or seasonaly without electricity stands around 1,100,000 people. More than 95 % of these people are low income people living in isolated areas of Mediterranean countries: Portugal, Italy, Greece, Spain and France.

Demonstration projects have clearly shown that photovoltaic systems and to a lesser extent WECS provide a technically acceptable solution for supplying these isolated areas with sufficient power to cover the basic domestic needs. Nevertheless, the investment cost for these systems is high and replication of the demonstration projects will be nil unless heavy financial support is available.

The financial support necessary to promote the use of Pv or WECS ranges from 90 % to 100 % of the total cost. This level is comparable to what happens for rural electrification through grid extension where public support usually covers 100 % of the cost. It is estimated that around 90,000 houses would be cheaper to electrify with Pv than through grid extension. The total power needed for such a programme would reach 50 MWp, costing 750 MECU.

1. SIZE OF NON ELECTRIFIED EUROPEAN POPULATION

There is to-day no reliable source of information covering the non electrified European population. In particular there is no agreement on definitions of what the non-electrified population is and how it should be evaluated.

When the figures exist and were obtained through known channels and methods, results should still be looked at with caution: whoever made the survey influences the results. Such a census can be conducted by users themselves, local politicians, electricity utilities, administration, development agency. Any of these people would have a tendency to over-or under-estimate their counts depending on their methods or interests. Censuses can never reach all people, especially those living far away, typically in non-electrified areas; finally, depending on its use, history, occupants, the same house may well be included in or excluded from the count.

From the literature and numerous interviews throughout Europe it was possible to obtain partial, more or less reliable data. From comparisons and deductions it was then possible to propose global estimates.

* Italy, Portugal and Spain conducted and published detailed and probably reliable data that provide figures for every region in those countries.

* In Greece, the administration and the electric utility agree on an estimate derived from the national census. Industry representatives convincingly explain that the official figure is very much underestimated.

* The official French data is based on no known detailed and systematic study. It is most probably underestimated.

* In Ireland and Great-Britain, administrations and electric utilities state that the non-electrified population is limited to a few thousand people, which is probably below but not far from the truth.

* In the rest of northern Europe, i.e. Belgium, Denmark, Germany, Luxembourg and the Netherlands, there is officially no non-electrified population although one admits that this answer comes from the fact that no one really worries much about rural electrification any more.

Table I shows the official figures together with EUROPE ENERGIE estimates for each EEC member state. Altogether, the European population living seasonally or permanently without electricity stands between 940,000 and 1,130,000, probably closer to the upper than to the lower limit. Southern Europe makes up for more than 95% of the total due to its combination of lower per capita income, rugged and poorly developed areas, low population density.

2. TYPOLOGY OF NON-ELECTRIFIED POPULATION

All available information indicate that the non-electrified population is made of two well separated groups.

* the first group (further refered to as Group I) consists of farmers, peasants, shepherds and fishermen living in southern Europe. They probably make up for more than 95% of the total non-electrified population.

* the second group (Group II), is made of people who occupy week-end or vacation houses as well as hunting, fishing, hiking, climbing huts or shelters.

characteristics and attitudes of Group I:

* occupation: nowadays, no modern activity can develop without electricity. People in Group I have traditional activity that they pursue in a traditional fashion. They do so because they have no

other choice, or because the other possibilities -moving to towns for example-are unacceptable to them.

* income: they generally have low incomes because their activity does not generate much wealth. They tend to keep their cash earnings saved to pay for basic investment related to their trade: fishing gear, farm machinery, diesel generator.

* age: younger generations tend to abandon such areas and move up to towns in order to get access to education, jobs and modern life. The average age of this group is high.

* evolution of group size: the number of people in Group I is decreasing for two reasons. Rural electrification programmes, although they may be slow, keep connecting more people to the grids and thus take them away from this group. Natural depopulation through older people's deaths and younger people migration amplifies the phenomenon.

* attitude toward Pv or wind systems: So far, people in this group have had a rather passive attitude; those who are now using such systems were provided them free of charge or at a very low cost through demonstration programmes. It should not be believed that people are always delighted to accept such systems because they are proposed free. Actually, people who live not too far from the grid, know that they hold good chances to be connected to it. They fear that by having benefited from subsidies for the installation of Pv or wind generators they may be dropped from the electrification priority list. Then also Pv or wind systems are viewed as no real substitutes for "true" electrification. By being connected to the grid they think they have access to an unlimited quantity of electricity whereas alternative systems have limits on their generating and storage capacity.

characteristics of Group II:

People in this group have in common that they occupy houses without electricity through a more or less deliberate choice. Houses like week-end or vacation houses, shelters or huts used for outdoor activities, are used by people that belong to middle or upper urban class. They have access to electricity at home all year round and accept or enjoy to do without during their leisures. They have financial resources sufficient to pay for a substantial portion of electrification costs should they wish to install electricity and being given the opportunity to do so.

3. POTENTIAL MARKET FOR PV SYSTEMS USED FOR RURAL ELECTRIFICATION

Knowing how much it costs to install a power line (7,000 to 30,000 ECU/km), it turns out that in terms of investment, Pv is economical for an isolated house as soon as it is located 800 m to 1.5 km away from the grid (pv power of 800 Wp). It was possible to define lower and upper limits on parameters allowing to calculate the total potential market in terms of users, houses, power capacity and investment assuming that the only competition is between grid connection and Pv. This calculation is shown on Table II. The minimum potential market stands to-day at 20 MWp. Probable figure is around twice as large but one must add that the progresses of traditional rural electrification programmes and the on-going depopulation of depressed areas work are slowly but regularly lowering this figure.

Heavy public support is the only possibility for this particular market ever to develop. No price decrease resulting from technology or productivity improvements would be sufficient to significantly alter the competition. For example, even a 50% drop in Pv systems price, a feat not readily achievable anytime soon, will still not allow a southern European farmer to buy such system, simply because he can neither spend 20,000 ECU nor 10,000 ECU on it. Actually price decreases will only have an effect on the global macroeconomics of this market, moving the border line between Pv and grid connection in favour of Pv.

4. LEGISLATION AND FINANCIAL PROCEDURES

In most of Europe, rural electrification was completed 10 to 20 years ago and electrification rates within the Community range from 96% to more than 99.9% of the total population. When an isolated house is connected to the grid, the investment is shared by the customer, the utility and public authority. The cost sharing system can be complex and the shares of the different contributors vary depending on the use of the house, income of user, special programme for the area, etc. To provide just some examples:

* in Greece, the nationalized electricity company PPC pays the full cost. Since funds are limited, electrification goes first to people that are on a priority list.

* in Ireland, the public electricity company ESB only pays for 60% of the cost, the remaining 40% being charged to the customer.

* in France, a special fund called FACE collects a tax on all urban electricity users. This fund and local authorities pays for the total cost.

Whatever the financial procedures that apply, in every European country where rural electrification is still going on at a significant pace, public support almost always reaches 100%. On the other hand, existing legislation is poorly -if at all- tailored to support electrification through pv or wind systems: in most cases procedures set up for traditional electrification are not applicable to alternative energy systems.

5. TECHNICAL MARKET

Electricity is needed in numerous non-electrified areas to power various types of technical installations: telecommunication relays, lighthouses, beacons, sensors etc. This market has some essential characteristics:

* highly professional: this is a professional market where purchasing decisions are made on essentially technical and economical grounds.

* single decision maker: for any given project, there is usually only one decision maker (for example port or lighthouse authority, TV company, Forestry department).

* single purchaser: purchases are made by one entity for its own use. Money comes from one single source.

* viable without subsidy: potential purchasers are perfectly capable of paying for the full price of the equipment and they often do. Of course, they will seldom miss the opportunity of a subsidy if they see one.

This sector grows on its own, at a rate dictated by technical needs and technological solutions available to satisfy them. The steady stream of sales that the sector generates constitutes the vital base load essential to manufacturers.

The most active part of this market so far is the sea light sector. There are around 12,000 navigation aids scattered along European shoreline (see Table IV). Wind generators have long been used to power buoys or small lighthouses and in the last five years, Pv systems have started being installed in increasing number, especially to power low power permanent or flashing lights which require a capacity of 10 to 80 Wp.

6. CONCLUSIONS AND RECOMMENDATIONS

* The potential for Pv rural electrification is strong but will only materialize if a comprehensive subsidy programme is set-up.

* A strong effort for Pv electrification cannot take place if proper legislation is not enacted. One of the most worrying problems to face is the probable rise of fraud and equipment theft along with the development of black market.

* The professional and vacation house markets do not need any subsidy. They go through slow, regular expansion but are nevertheless limited.

* Thanks mostly to demonstration projects, the viability of Pv and wind systems is now well established: further efforts should turn preferably to:

*** improve standardization of components so that they become interchangeable and available from competing manufacturers.

*** decrease production costs through the lengthening of production series thanks to larger projects.

*** improve components reliability through systematic endurance testing and imposed longer warranties on equipment.

*** improve marketing channels so that components and complete systems become available from outlets such as large hardware stores or agricultural/fishing equipment retailers.

Table I: Size of non-electrified populations
(total count and percentage of total)

EEC Member States	official figure	EE assessment
BELGIUM	-	500 (-)
DENMARK	-	500 (-)
F.R.G.	-	2,000 (-)
FRANCE	10,000 (-)	50,000 (0.10%)
GREECE	27,000 (0.27%)	150,000 (1.50%)
IRELAND	2,000 (0.06%)	5,000 (0.15%)
ITALY	450,000 (0.80%)	500,000 (0.90%)
LUXEMBOURG	-	-
NETHERLANDS	- (-)	500 (-)
PORTUGAL	300,000 (3.70%)	400,000 (4.90%)
SPAIN	150,000 (0.40%)	200,000 (0.50%)
UNITED KINGDOM	1,000 (-)	3,000 (-)
TOTAL	940,000	1,131,500

Table II: Global potential market for pv systems used for rural electrification

ESTIMATES :	LOW	MEDIUM	HIGH
1. Total non-electrified population (including seasonally occupied vacation houses).	1,000,000	1,200,000	1,500,000
2. Average number of occupants per house.	5	4	3
3. Proportion of houses for which grid connection is more expensive than alternative energy systems.	20%	30%	40%
4. Number of houses to electrify with alternative systems.	40,000	90,000	200,000
5. Percentage of houses for which Pv is the best alternative.	75%	80%	90%
6. Percent of houses requiring 800 W (others supposed to need 200 Wp) Average power per house.	70% 620 Wp	80% 680 W	90% 740 Wp
7. Total solar cell power needed.	19 MWp	49 MWp	133 MWp
8. Total investment (15 ECU/Wp).	285 MECU	735 MECU	2000 MECU

Table III: Sea lights along European shores (1985)

Member States	lights on fixe structures	lighted buoys	radio beacons	radar beacons
BELGIUM	74	111	4	-
DENMARK	296	372	25	25
F.R.G.	1,661	677	14	15
FRANCE	1,797	1,044	48	22
GREECE	696	95	2	2
IRELAND	97	104	17	10
ITALY	1,098	101	19	2
LUXEMBOURG	-	-	-	-
NETHERLANDS	483	567	10	10
PORTUGAL	327	109	8	1
SPAIN	921	186	27	4
UNITED KINGDOM	360	699	66	67
TOTAL	7,810	4,065	240	103

Contract No: XVII/7032/85/CDC/402.5

DC APPLIANCES FOR USE WITH PV GENERATORS

W B Gillett and J E Bates
Sir William Halcrow & Partners Ltd
Burderop Park, Swindon, SN4 OQD, UK

SUMMARY

A preliminary survey of the dc appliances, which are currently available in Europe, is reported. More than 80 companies were contacted by post in the spring of 1987, and asked for prices and technical details of the dc powered appliances which they could supply.

It was found that the leisure industry has the most well established marketing network for dc appliances. However, these are sold for use in boats, caravans and other vehicles where energy efficiency is not considered to be of primary importance. The range of energy efficiencies is therefore rather wide and many appliances are not marked with either their rating or their daily energy consumption.

With the exception of dc lighting, where the market is well developed, the consumer has a very limited choice of dc appliances when compared with his choice of ac appliances.

Most of the dc appliances appeared to have been developed by relatively small companies. It is therefore concluded that there may be many more small local suppliers in the Community which were not contacted during the survey. It is recommended that a more detailed survey of the appliances available in each country should be carried out.

1. BACKGROUND

The CEC photovoltaic demonstration programme involves a number of projects where small pv generators are used to electrify rural houses, farms, hostels and mountain refuges, which are sited in remote locations far from the electricity grid. In these projects, the pv generators typically provide power for lighting and small domestic appliances such as refrigerators, televisions and kitchen equipment.

PV generators may be connected to an inverter and used to provide an ac supply. This allows ordinary ac appliances to be used, but the overall efficiency of the generating system is reduced because of losses in the inverter. Also the overall cost of the generator may be substantially increased by the cost of an inverter especially in the case of small pv systems.

An interesting alternative is to install dc appliances which may be run from a low voltage (12V or 24V) power supply thus removing the need for an inverter. Dc appliances are attractive because the pv system can be both cheaper and more efficient, but they are generally more expensive than equivalent ac appliances, and are more difficult to obtain.

In order to investigate the availability, choice, cost and efficiency of dc appliances in Europe, a survey was carried out in the spring of 1987. In this paper, the preliminary results obtained from the survey are presented.

2 THE SURVEY

A total of 82 companies in Europe, the USA and Canada were contacted. These included:

(1) The PV Industry
- PV module manufacturers
- Companies which design and supply complete pv systems

(2) The Leisure Industry
Companies manufacturing or supplying dc equipment for use in the leisure market, ie for boats, caravans and other vehicles and for camping

(3) The Electrical Equipment Industry
- Major manufacturers of domestic electrical products
- Wholesalers of electrical products

The Companies were asked for technical and commercial data for each dc appliance which they supplied, including the electrical rating of the appliance and the retail price. American and Canadian companies were also asked for the names and locations of their European distributors.

A relatively large number of UK companies was approached, using trade directories as a means to identify companies which might supply dc appliances. The British camping and caravanning organisations were also asked to supply lists of dc equipment manufacturers and distributors.

In general the level of response to the survey was disappointing; under half of the companies replied and of these only 20 companies were actually able to supply dc appliances.

Several of the UK companies which did not respond initially were contacted again by telephone. The discussions held by telephone were interesting, because they provided an opportunity to explore the problems involved in obtaining information on dc appliances.

3. RESULTS OF THE SURVEY

3.1 General

The information received from companies varied greatly in quality, and more than half of those contacted did not reply at all. The follow-up contacts which were made by telephone revealed that ordinary consumers might find it quite difficult to obtain both commercial and technical information on dc appliances.

In the electrical equipment industry, the marketing departments were generally unable to provide information on dc appliances, and referred the matter to technical departments. In fact, very few of the main electrical companies appeared to have any interest in dc appliances.

In the leisure industry, the marketing departments were much better informed and were generally able to offer a range of dc appliances together with detailed price data. However, they were in general not able to provide technical data concerning the operating efficiencies and specifications of the appliances.

The pv industry response was most disappointing and suggests that the pv companies are not yet able to offer appliances for use with their pv generators.

Data from several companies was already available to DG XVII and Halcrow through previous contacts. Although some of these companies did not respond to the survey, their information was included in the data analysis.

In order to permit comparisons between products from different countries, all of the quoted retail prices were converted to ECU, using April 1987 rates of exchange.

3.2 Lighting

Information on low voltage dc lighting was received from 18 companies, 13 in Europe and 5 in the USA. Four types of lighting were reported to be available for use with low voltage dc power:

- Incandescent lamps
- Tungsten halogen lamps
- Fluorescent lighting
- Low pressure sodium lighting

The efficiency of a lamp is usually presented as the ratio of the light output of the lamp, measured in lumens to the power consumption of the lamp. Typical values for the efficiencies of different types of lighting are given in Table I.

Lighting Type	Range Available			Price (ECU)	
	Watts	Lumens	Lumens/Watts	Complete Lamp	Replacement Bulb
Incandescent	5-100	54-1800	10 - 18	9 - 20	0.4 - 3.1
Tungsten or Quartz halogen	5-75	400-1800	23 - 35	25 - 70	6.5 - 11.5
Fluorescent	6-20 20-80	400-6300	50 - 80	15 - 40 30 - 70	3.3 - 10
Low pressure sodium	18-35	1600-4800	90 - 140	130 - 340	15 - 20

Table I - Range of Low Voltage DC Lighting Available in Europe and USA

The cheapest and least efficient type of lighting is the incandescent tungsten lamp which is commonly used for domestic lighting. Tungsten or quartz halogen incandescent lighting can be twice as efficient and the bulbs typically have a lifetime of which is twice that of ordinary incandescent bulbs.

Tungsten or quartz halogen bulbs are often used in spot lights or for downlighting and there appears to be a growing market for this type of lamp in shop display lighting.

Low voltage fluorescent lights employ a small, high frequency inverter in the ballast of the luminaire. This produces a light with less flicker than fluorescent lights running at 50 Hz from the mains. Fluorescent lighting is available in a variety of styles, including long tubes, circular tubes, and lamps which will fit into an ordinary bulb holder. The range is well developed as there is already a large market for low voltage lighting in lorries, coaches, ambulances, and other vehicles as well as caravans and boats.

Low pressure sodium lamps are extremely efficient and have a long life. However they produce a monochromatic yellow/orange light which limits their use to applications such as street lighting, where the quality of light is not important.

The price of each type of lighting, varies according to the wattage of the lamp and the type of mounting or fixing. The efficiencies of each of the main lamp types together with their typical replacement bulb prices are shown in Figure 1, where it can be seen that sodium lamps are the most efficient and conventional incandescent bulbs are the least efficient. Flourescent lamps offer a reasonable compromise for most pv powered applications other than streetlighting.

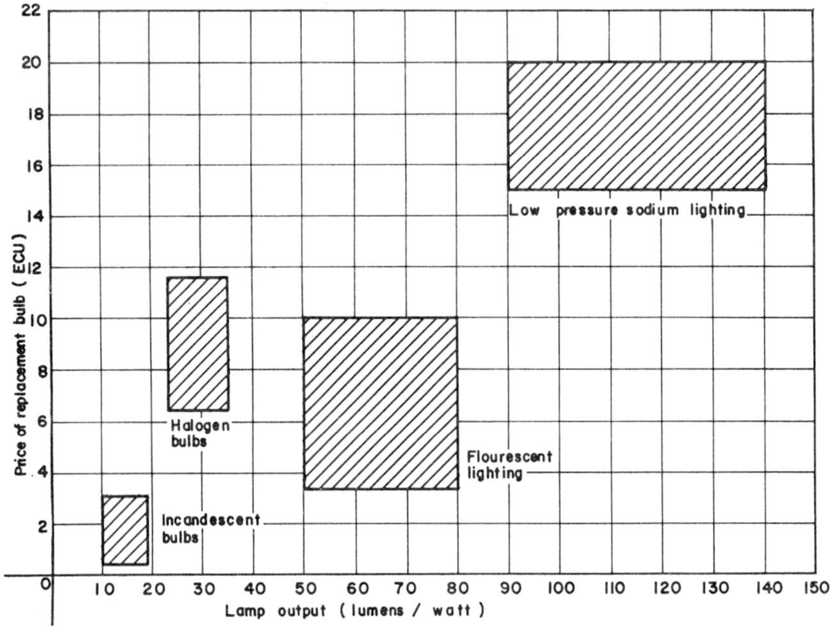

Figure I <u>Typical lamp outputs and replacement bulb prices for different types or lighting</u>

3.3 Refrigerators

Fifteen companies supplying dc powered refrigerators and refrigerator/ freezers were identified during the survey: 10 in Europe and 5 in USA. Their products include medical refrigerator/freezers designed for transporting and storing vaccine, portable cooler/chiller boxes, small refrigerators and refrigerator/freezers designed for use in boats and caravans, and refrigerator/freezers and deep freezers of a similar size to those normally found in houses. All of the products identified during the survey run at 12 volts and many could also operate at 24 volts. Several models had the additional options of running from a mains supply at 110V or 240V or from liquid propane gas.

The refrigerators employ either an absorption cycle or a compression cycle. The small refrigerators which employ an absorption cycle require only an external heat source and no pump. They can therefore easily be adapted to a 12V power supply and require little maintenance. The absorption cycle is however less efficient than the compression cycle. Typically the heating element of an absorption cycle refrigerator is rated at 70-100W, giving a running current of 6A to 8A. The operating time of the element depends on the cooling load.

Most refrigerators which employ compression cycles require a high starting current which is usually at least 3 times the running current. Depending on the size of the refrigerator, typical running currents lie in the range of 2A to 4A, and the running time is typically 30-40% of the operating period.

As the two types of refrigerator cannot easily be compared in terms of running current, the average energy consumption over 24 hours with an outside ambient air temperature of 32°C has been used to evaluate the relative efficiencies of pv powered refrigerators. The available data for refrigerators of different internal capacities is shown in Figure 2 as a function of retail price. In general it can be seen from this figure that the more efficient refrigerators are also more expensive.

The cheaper dc refrigerators are generally based on mass produced conventional models which have been converted to run with a dc power supply. Several of the more expensive refrigerators have been specially designed with higher efficiency compressors and better standards of insulation than would be found in conventional models.

Prices vary between the different manufacturers, and there is a wide range of operating efficiencies. However, because so few manufacturers publish data on energy consumption it is not easy for the consumer to choose the most cost effective refrigerator for use with a pv generator.

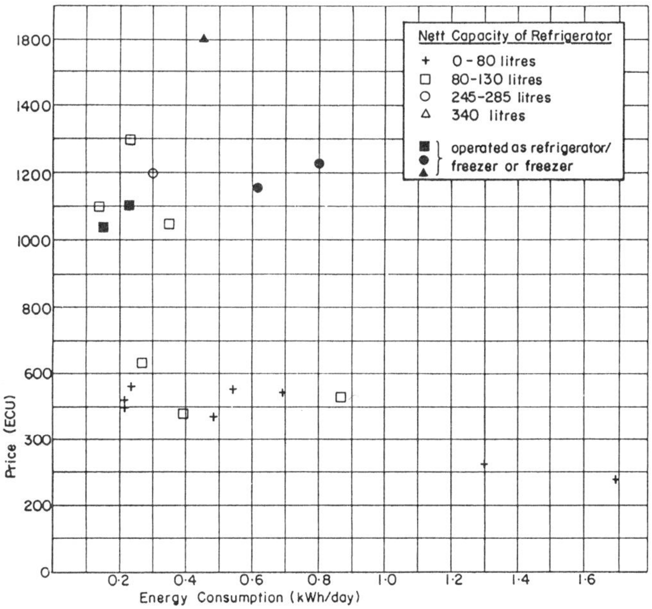

Figure II Energy consumption of refrigerators and refrigerator freezers over a 24 hour period as a function of retail price

3.4 Televisions

No responses were received from any of the television manufacturers contacted. However contacts with local television suppliers showed that small televisions, which could be run from a low voltage power supply were easily obtainable in the UK. Black and white televisions with 250 mm or 300 mm screens are available for between 70 ECU and 100 ECU. These can be used directly with a 12V dc power supply. Colour televisions with screens of between 250 mm and 375 mm vary in price between 280 and 425 ECU, but are usually sold only for use with mains power. Most colour televisions need to be fitted with an adapter, which plugs into the back of the set when they are used with a 12V dc power supply. Small black and white televisions are typically rated at 12-20W and small colour televisions at 35-50W when used with dc power.

3.5 Other dc Appliances

Other dc appliances for which data were obtained during the survey are listed in Table II. Most of the data for these appliances were obtained from the catalogues and brochures issued by large UK distributors of boating and caravan equipment and by distributors in the USA which specialise in supplying equipment for use with pv systems or wind generators.

The survey did not attempt to cover the wide range of portable battery powered devices which are currently available, such as radios, cassette players, hand held whisks, blenders and other kitchen appliances. All such appliances could be run either directly from a low voltage dc supply or by using batteries which are recharged using a dc supply.

APPLIANCE	No of Suppliers	Rating (W)	Price (ECU)
Toaster	1	180	41
Foodblender	3	100	45-65
Extractor Unit for Cooker	1	?	60
Iron	1	?	15
Vacuum Cleaner	2	120	27-30
Hair dryer	2	180	7-10
Hair styler	2	25	5-10
Record player turntable	1	6	130
Slide projector	1	?	174
Portable typewriter	1	?	313
Soldering iron	2	?	5-6
Electric fence	3	?	70
Fan	5	6-18	20-65
TV satelite dish	1	15	911

Table II - Miscellaneous low voltage dc appliances identified during the survey

4 DISCUSSION

It was clear from the survey that dc appliances are in general more difficult to obtain than conventional ac appliances, and that the consumer who wishes to employ dc power cannot at present enjoy the same choice of appliances as he would if he were to adopt an ac power supply. Moreover, many of the staff who were contacted during the survey in conventional electrical supply companies did not appear to understand the difference betwen dc and ac powered appliances.

The product range which appeared to be the most well developed for use with pv power was lighting. The general increase in consumer demand for energy efficient lighting has encouraged the development of a wide range of efficient low voltage dc fittings, and these are now available from many sources. Small black and white televisions, which are suitable for use with dc power, are also widely available.

The leisure market has encouraged the development of a range of dc appliances for use in boats, caravans and vehicles, but energy efficiency does not appear to have been of primary importance in these developments. Apparently similar appliances which are offered on the leisure market, are currently sold with widely different energy efficiencies. It is concluded that, whilst this sector of the dc appliance industry has the most strongly established marketing network, it is weak in its encouragement of energy efficiency. Perhaps the development of more energy efficient dc appliances for use with pv generators would be encouraged if all electrical products were to be marked with details of their daily mean energy consumption and power rating.

Whilst it was apparent from the survey that there is some international trade in dc appliances, a large proportion of the dc appliances, which were identified during the survey, were produced by small specialist suppliers. Such suppliers are not easily located from a distance, so it is likely that there are many more small suppliers of dc equipment, which were not contacted during the survey, in each country of the Community.

5 CONCLUSIONS

The preliminary survey reported here should be considered only as indicative. A more detailed study of the dc appliances available from suppliers in the Community should be carried out to confirm the situation in each country.

Many dc appliances are sold to the leisure industry, but a substantial proportion of these have a poor energy efficiency. To encourage the development of more energy efficient dc appliances for use with pv generators, a labelling scheme should be introduced to ensure that all appliances are marked with their electrical power rating and average daily energy consumption.

5 POINTS ARISING DURING DISCUSSIONS

Several discussion sessions were held during the meeting, and the following important points were raised.

5.1 Lightning Protection

Most systems need to be protected to some extent against lightning. Mr Blaesser from the JRC pointed out that some protection is given by the equipotentiality of structures, ie all structures and parts of the system should be connected to a common earth. Array structures can also be protected by conventional lightning poles and wires. For a central PV plant with an extended grid, it is also necessary to use special devices to protect the grid from over-voltages induced by lightning strikes nearby. This can only be achieved by using over-voltage protection devices which react very quickly (in nano seconds). Inverter failures are often caused by over-voltages.

Monitoring systems should also be protected from over-voltages. Data loggers are generally equipped with over-voltage protection, but over-voltage protection may need to be built into data acquisition systems based on microcomputers.

5.2 Choice of ac or dc Systems for Dwellings

Maximum efficiency is obtained from a PV system if it is used to produce dc power. Ac systems require the use of inverters, which typically have a maximum efficiency of about 90%. However, appliances for use with a dc power supply are often more difficult to acquire and more expensive than equivalent ac appliances. They are mainly sold on the leisure market and are less efficient and of lower standard than commercial appliances. Mr Aubert reported that in his experience it may be psychologically important for some residents of PV electrified dwellings, to be able to buy and use the same standard appliances as those residents who are connected to the grid. However, in France the available dc appliances are often technically inferior to equivalent ac appliances, and this discourages consumers from choosing dc appliances.

Mr Blaesser commented that it is not normally cost effective for dc and ac systems to co-exist in the same house. Therefore where there are already ac appliances which cannot easily be replaced by an equivalent dc appliance, an ac system should be installed.

5.3 System Utilization

System utilization appears to be heavily dependent upon the lifestyle and economic standing of the system users. Users with low incomes may not be able to afford to purchase new electric appliances very rapidly. Therefore their PV systems will initially appear to be overdesigned.

Also, many rural households are habitually very economical in their energy consumption and will use the system very carefully.

It was agreed that more experience was needed in order to improve our ability to size PV systems for rural dwellings and that there was unlikely to be a standard power requirement for a family of a certain size. Mr Dichler suggested that it might be possible to use the consumption figures used by electricity utilities for planning purposes. Mr Blaesser pointed out that monitoring results will provide data on system utilization which may assist with future system designs.

5.4 Inverters for Grid Connected Systems

For PV systems which are connected to the grid, either a line commutated or a self commutated inverter may be used in the system. Although self commutated inverters produce a high quality sine wave, line commutated inverters are intrinsically safer because they automatically close down when the main grid fails. For electricity utilities, safety is of the highest priority, so it is likely that they will prefer the use of line commutated inverters. It was suggested that experience could be gained by looking at how other types of generator are connected to the grid.

5.5 Inverter Manufacturers and Inverter Technology

There are many manufacturers producing inverters which range in size from 10W to 100 kW but many of these inverters are intended for the leisure and hobby markets. Of the commercial inverters on the market only a few are suitable for use with PV systems.

Typically each PV system requires a unique inverter which has to be specially designed or specially modified from a standard inverter. It is unlikely therefore that a manufacturer would wish to produce a range of standard PV inverters.

Several different techniques are currently used in inverters, for PV applications and there is not yet enough experience to show which of these techniques produces the best results.

5.6 Stand Alone PV Generators and Central PV Generators

The advantages and disadvantages of electrification by small stand alone PV systems or central PV plants were discussed. Several social problems could be caused by centralized PV plants, for example it might be necessary for the power consumption of individual users to be limited, but it was not clear how this should be achieved. Stand alone systems encourage users to recognize the limits of their PV system and adjust their energy consumption accordingly. A suggested method of limiting the power consumption of a user was to install fuses so that only appliances with low power consumption could be used.

It was noted that the price of electricity charged by a utility is normally the same for all users, regardless of how the electricity is generated, ie utilities may not charge more for electricity from a PV generator than for electricity from the main grid.

However, electricity utilities may not be able to afford the administration and maintenance costs associated with many small stand alone PV systems. They would therefore prefer to give these responsibilities to the system user.

5.7 Durability of PV Modules in a Marine Environment

Mr Dichler asked whether the PV modules used in lighthouse projects needed to be specially adapted to resist the humidity and salinity of a marine environment. Mr Kadella from Interatom reported that a special type of module with a stainless steel frame and specially designed connectors had been used in the Methoni and Lithari lighthouse projects. The representatives from AEG stated that all AEG modules have a glass front and back and a stainless steel frame, and so they are suitable for use in a marine environment. Mr Fonzi noted that for PV systems to be fully accepted by national and international committees on lighthouses the PV modules should be approved to Marine standards. It was suggested that the JRC at Ispra, which currently certifies PV modules might like to contact the relevant Marine authorities and determine whether additional tests are required to meet the marine standards.

5.8 Commercialisation of PV systems for Lighthouses

It was noted that the use of PV generators in buoys and navigation lights was now an accepted commercial application for PV which did not require further demonstration. However, all three of the demonstration projects discussed were on a much larger scale than current commercial applications. The hope was expressed that if these demonstration projects are successful, then they will help to raise confidence in the use of larger PV systems to power full sized lighthouses in many parts of the world.

5.9 Model Economic Calculation for Co-Generation Systems

In the discussion following the presentation on a model calculation for the economic assessment of PV systems, the question of how to treat grid connected systems and co-generation systems in the calculation was raised. In some systems it will be possible to consider the PV generator as a separate generator and thus use the existing model calculation. In other systems the same control equipment may be used by more than one generator, and the cost of the control equipment should therefore be apportioned between the different generators. As there are several co-generation systems and grid connected systems in the demonstation programme, it was agreed that the model calculation should be extended to cover such cases.

6 CONCLUSIONS

It was concluded that the meeting had provided an important opportunity for the contractors to exchange information and experiences, and that this would be particularly helpful to those who were about to embark on new projects.

The demonstration programme was shown to be progressing well and many monitoring results were reported to be becoming available. These results will form an increasingly important part of the publications about the PV demonstration programme which must be produced if its goals are to be met.

The PV Plant Monitoring Group was shown to be making an important contribution to the work of the demonstration programme. In particular, all of the projects will now follow the Monitoring Guidelines which have been developed with the help of this Group.

For large systems, which employ analytical monitoring it was concluded that a simple global monitoring system should also be installed to provide evidence that the PV plant has continued working if the analytical data acquisition system fails.

The aim of the demonstration programme is to assist the European PV industry by demonstrating the commercial viability of PV generators and PV systems in appropriate applications. The applications currently being demonstrated in the programme include, housing, lighthouses, warning and monitoring systems, telecommunication systems, pumping systems, and nature reserves. The products being demonstrated include PV modules, module support structures, batteries and battery housing, inverters, control equipment and end use devices.

This Contractors meeting provided an additional opportunity for the European PV industry to benefit from the demonstration projects. Information was exchanged on how systems should be sized, what lightning and over-voltage protection is needed for systems and on the typical utilisation of small systems in remote houses.

GUIDELINES FOR PV SYSTEM MONITORING

GUIDELINES FOR PV SYSTEM MONITORING

1. INTRODUCTION

Data obtained during the operation of a photovoltaic power plant are of interest for judging whether the design goals have been met. They are also of importance for a general assessment of the potential of PV-technology, and thus for the future use of such installations. Finally, they should provide a data base for the development of theoretical models for PV system design.

To obtain sufficiently complete sets of data for such purposes requires rather detailed or "analytical" monitoring in accordance with the reqirements set out in these Guidelines. In general, unless otherwise agreed, analytical monitoring will be applied for at least 24 months for all PV installations with peak power 5 kW or more.

In view of the cost involved in analytical monitoring, sufficient evidence for a general assessment of the performance of a PV installation may be obtained by determining the overall energy balance from regular readings of suitable counters. This procedure, referred to as "global" monitoring, will in general be applied for all PV installations with peak power less than 5 kW.

For some small installations, however, it may be appropriate to apply analytical monitoring for a few months only and then revert to global monitoring for the remainder of the monitoring period.

The treatment of the large amount of information collected for monitoring purposes, with the aim of making reliable comparisons between different stations, requires the adoption of a commonly agreed minimum set of quantities to be measured as well as standard recording techniques. These guidelines give the information which is necessary for the implementation of such monitoring systems.

2. REQUIREMENTS FOR ANALYTICAL MONITORING

2.1 Parameters to be Recorded

a) Main System

An automatic data acquisition system is required for analytical monitoring. This may be achieved by a dedicated computer with suitable peripherals and programs or a data logger capable of recording the data on the required medium in the correct format, as shown schematically in Fig.1. Recommendations for the selection of computer systems and data acquisition equipment are given in Annex A.

The minimum set of parameters to be monitored is listed in Table I, with symbols and units. It may be necessary to record additional parameters according to the requirements of a particular project. The required accuracies are stated in the footnote to this table. Apart from the meteorological data (including temperatures) all parameters are currents, voltages or energies characterising the operation of the system (consisting of the PV field and the associated loads, power conditioners, batteries etc.) during the standard time interval. The points of measurement are indicated in Fig.2.

In general, only aggregate values of each parameter are required: for example, if the energies of subfields of the total array field are separately measured, only their sums should enter the final monitoring record as JA,UA; if several DC-loads are present, the total current to all these loads gives JLDC; if the DC/AC interface in Fig.2 consists not only of an inverter but possibly also of a separate rectifier which charges the battery and supplies the DC-loads from a public grid when PV-power is lacking, its contribution should be included (with a negative sign) in JII and EIO.

The ac power measurements may be performed directly by analog techniques and the resulting power signal digitized by the A/D-interface of the data acquisition system (computer or data logger).

All numerical data should be hourly sums or averages of measurements taken at 1 minute intervals. However, if such averaging is not possible, instantaneous samples taken at 10 minute intervals are acceptable.

b) Back-up System

In addition to the main data acquisition system, an independent system with mechanical counters should be installed, to record the solar input by calibrated reference cell (IGc), useful electricity output (kWh or Ah) and product flow, if applicable. (A separate record of IG need not be made provided the Contractor demonstrates that the records from the nearest meteorological station are sufficiently representative of conditions at the site.)

These counters must continue to operate even when the main data acquisition system is out of action for any reason. The cumulative readings of these counter devices should be recorded manually at monthly intervals and the data sent to the JRC together with the other data records. (See 2.3 below).

2.2 Format of Records

The data should be recorded or copied off-line in the format given below on one of the following media:

(a) an ECMA 46 magnetic tape cassette
or
(b) an IBM-compatible magnetic tape
(9-track, 1600 bpi)
or
(d) a MS-DOS (IBM PC-DOS compatible) 5.25" and 3.5" floppy disk.

Experience has shown that ECMA-46 cassette equipment is less sensitive to environmental stresses than floppy disk and tape drives; commercial data loggers are available with this standard. In most cases data recording on floppy disk via a PC running under MS-DOS may be the most economic choice.

Under special circumstances, other recording media or direct transfer of data by satellite or telephone may be accepted subject to agreement between the parties concerned. For some installations of particular interest, the JRC may provide its own data acquisition systems for a limited period.

The results of a measurement period (1 hour if averages over one-minute samples are taken, or 10 minutes if no averaging is used) form a data group. Each data group is subdivided into records, comparable to lines in printing. Records are separated by a carriage return ('CR'=ASCII(13)). Each record is built up of individual fields, separated by a comma. The record length must not exceed 72 characters.

All data are written in ASCII code. Since the comma (","=ASCII(44)) is used as a field separator, all data can be presented in free-field format without the necessity of keeping leading zeros or spaces, as signed integers or decimal fractions, with the stop ("."=ASCII(46)) used as the decimal point. So-called scientific notation (E-notation) is not permitted. If any field in the format of a record (given below) is not applicable or data are not available for it, it has to be indicated as an empty field by its final comma following directly the preceding one. However, all commas directly preceding the end-of-record should be suppressed. Thus, if in data record 2 only the 4th and the 6th item is available, that record should be written as:

2,,,,item 4,,item 6'CR'

To provide a maximum of flexibility for the each installation to record additional data of its own choice, the number of records per data group may vary. Each data group starts with a header record of the following form:

%Station%Date,Time,Comments

where "Station" is the name of the site, of which only the first four letters are mandatory (full length is permitted), Date specifies the day of the measurement in the form dd-mm-yy (here leading zeros must be written), and Time the moment of recording in the form hh.min. The time of recording is the time at the END of the hour in which measurements were taken, or the instant of recording if no averaging is used. (Time always refers to the true zone time, not to legal or daylight saving time).

In the case of hourly data recording, the space allocated for comments may be used to record the times of switching events (e.g. switching from load to battery or switching in parallel inverters) in a code devised by the installation manager. Other comments may be included at the discretion of the installation manager. Thus two examples of headers are:

%NICE%03-05-84,15.33

%FOTA%25-03-85,13.15,EXTENDED RUN

The header record is followed by up to 9 data records, distinguished by their record number, which is the first item of each record. The symbols used are those given in Table I.

Data record 1 contains the meteorological data and those of the PV field - including the battery backup - while data record 2 shows the energy balance of the consumer; data records 3 to 9 are at the free disposal of the installations (or may be absent). The data in records 3 to 9 is considered proprietary information which will not be recorded in the PV-data bank of the European Communities nor appear in the periodic monitoring reports.

DATA RECORD 1:

1,IG,IA,TA,UA,JA,JC,JBI,JBO,UDC

DATA RECORD 2:

2,JLDC,JII,EIO,ELAC,ER,EG

2.3 Submission of Data to the JRC

All recordings (covering periods not less than one month and not more than three months) should be sent to the JRC Ispra together with a note giving information helpful for the interpretation and evaluation of the data. The note shall cover such matters as:

- Any faults or accidents
- Any maintenance operations, such as cleaning of PV arrays
- Any changes to system parameters, such as the change of array tilt angle
- Data acquisition system: recalibration of sensors or any down time
- Energy to load: any unusual conditions
- Weather: any extreme events (e.g. hail, snow)
- Readings of back-up counters for IGc and kWh (or Ah)

A suggested pro-forma for the covering note is given in Annex B.

In addition, with the first set of records (if not previously submitted), a data sheet setting out basic information relating to the plant should be sent to the JRC. A suggested pro-forma for the data sheet is given in Annex D.

2.4 Data Processing at the JRC

The data received by the JRC will be entered into the main computer of the European Solar Test Installation (ESTI) for analysis and evaluation, including the preparation of a report summarising the main results for that period; the files will be stored at ESTI on tape for backup, while another copy is transmitted via the local computer network of the JRC to the mainframe computer of the ISPRA Computing Centre. Since the Computing Centre is connected to the Italian public network (ITAPAC) which is linked with public data networks in the other member states, these data are accessible in principle to all authorized users within the EC.

The original recording is returned to the station together with the summary report.

3. REQUIREMENTS FOR GLOBAL MONITORING

3.1 Parameters to be Recorded

A manual system of recording data is appropriate for global monitoring. The minimum set of parameters to be monitored is listed in Table II, with symbols and units. The required accuracies are stated in the footnote to the table.

The cumulative readings of the listed parameters should be recorded manually at intervals not exceeding one month and preferably much shorter (daily or weekly if possible).

3.2 Format of Records

The cumulative data should be entered onto standard record sheets. A suggested format is given in Annex C.

3.3 Submission of Data to the JRC

Copies of the record sheets covering a period of not less than one month and not more than three months should be sent to the JRC Ispra together with a note giving information helpful for the interpretation and evaluation of the data. The note shall cover such matters as:

- Any faults or accidents
- Any maintenance operations, such as cleaning of the PV array
- Any change of system parameters, such as a change of array tilt angle
- Data acquisition system: recalibration of sensors or any down time
- Energy to load: any unusual conditions
- Weather: any extreme events (e.g. hail, snow)

A suggested pro-forma for the covering note is given in Annex B.

In addition, with the first set of records (if not previously submitted), a data sheet setting out basic information relating to the plant should be sent to the JRC. A suggested pro-forma for the data sheet is given in Annex D.

3.4 Data Processing at the JRC

The data received by the JRC will be entered into the main computer of the European Solar Test Installation (ESTI) for analysis and evaluation. A report summarising the main energy balance results for the period will be produced and a copy sent to the manager of the PV installation concerned.

TABLE I
Recorded Parameters for Analytical Monitoring

Parameter	Symbol	Units
Irradiation, global (horizontal)	IG	kWh/sqm
Irradiation, array plane	IA	kWh/sqm
Ambient temperature in the shade	TA	deg.C
Array output voltage	UA	V
Array output current (total)	JA	A
Converter output current	JC	A
Current input to battery	JBI	A
Current output from battery	JBO	A
DC line voltage (Battery voltage)	UDC	V
Current to all DC loads	JLDC	A
Inverter/rectifier DC current (+/-)	JII	A
Inverter/rectifier AC energy (+/-)	EIO	kWh
Energy to all AC loads	ELAC	kWh
Energy to(+) and from (-) grid	ER	kWh
Energy from auxiliary generator (e.g. Diesel, wind turbine etc.)	EG	kWh

Notes:

1. IG should be measured with a pyranometer or a calibrated solar cell. The electrical data shall be determined to within +/- 2% f.s.

2. A separate record of IG need not be made provided the Contractor demonstrates that the records from another recording station are representative of site conditions.

3. ELDC may be recorded as Ah if kWh recording is not appropriate.

TABLE II
Recorded Parameters for Global Monitoring

Parameter	Symbol	Units
Irradiation, global (horizontal)	IG	kWh/sqm
Array output current (total)	JA	A
Energy to all DC loads	ELDC	kWh
Energy to all AC loads	ELAC	kWh
Energy to(+) and from (-) grid	ER	kWh
Energy from auxiliary generator (e.g. Diesel, wind turbine etc.)	EG	kWh
Hours load connected	TAV	hrs

Notes:

1. IG should be measured with a Class 1 pyranometer, IA by means of a pyranometer or a reference device; TA should be determined to within +/- 2 deg.C and the electrical data to within +/- 2% f.s.

2. The energy consumed by the control and data acquisition systems should be considered as part of the external load and not as internal losses (i.e. included in JLDC or ELAC). If possible, a a separate measurement of internal losses should be made. If not practicable, an estimate of losses should be made and entered on the data sheet (Annex D).

3. Irradiation in the array plane is understood to be equivalent irradiation if a reference device is used. If a pyranometer is used for IA, certain comparisons between plants cannot be made, due to spectral mismatch.

Fig. 1. General Layout of the Data Monitoring System for Analytical Monitoring

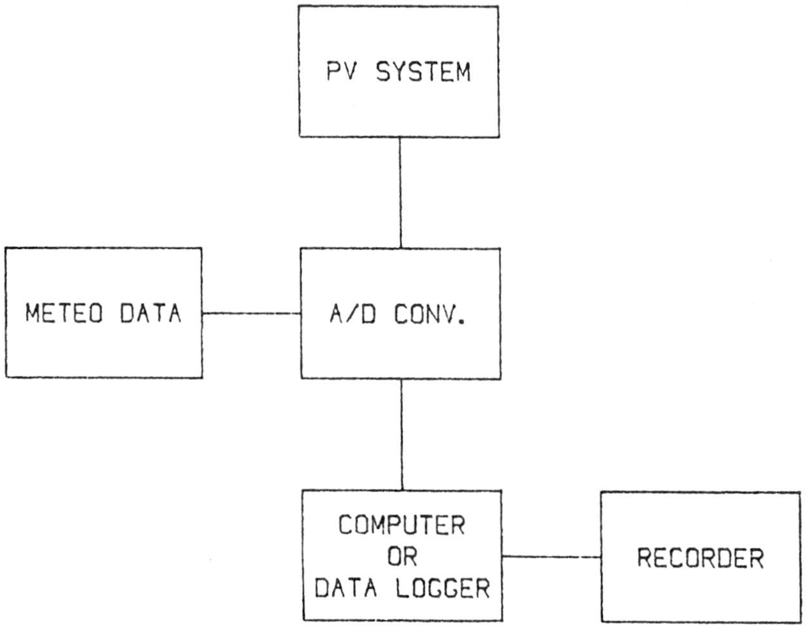

Fig. 2. Energy Flow Diagram

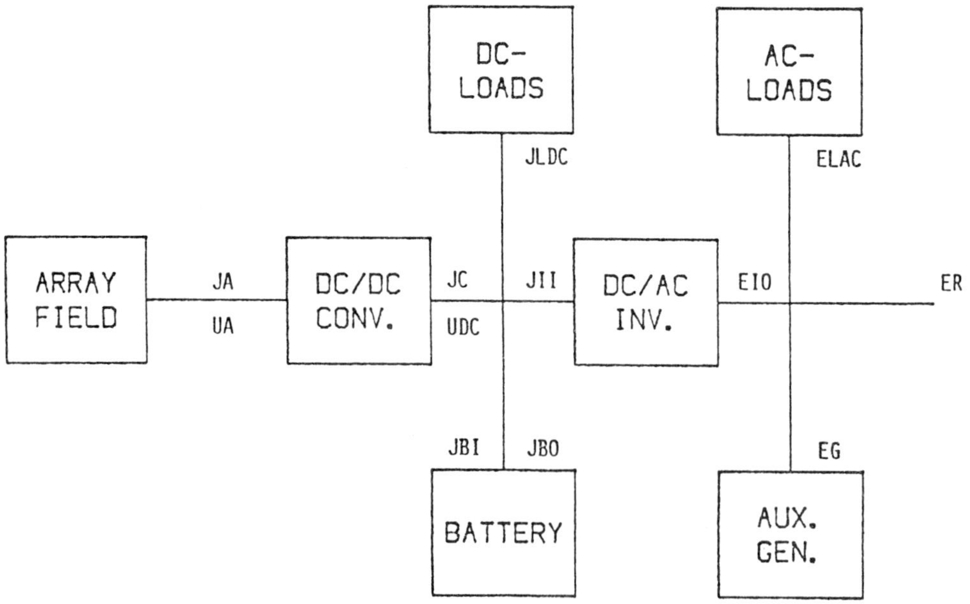

ANNEX A

RECOMMENDATIONS FOR THE SELECTION OF COMPUTER SYSTEMS AND DATA ACQUISITION EQUIPMENT

The following recommendations relate to computer-based control and/or data acquisition systems and are based on experience gained from the operation of existing photovoltaic plants.

1. The computer system should be easy to use.

2. Technical support for the hardware and software should be available and this should not be limited to individual suppliers.

3. The hardware and software should all be properly documented. In the case of software, there should be a user's manual and a programmer's manual. The purpose of the documentation is to reduce the problems associated with changes in personnel and to permit improvements in system design.

4. The software should be written in a recognised high level language (e.g. FORTRAN, C, PASCAL, etc). This will ensure that future changes in software are not restricted to individuals or a specific software supplier. The computer system should be programmable on site, either with a dedicated computer terminal or a 'lap-top' portable computer.

5. An inventory of spare parts should be kept on site. Standard cards should be available as spares. A list of suppliers of cards and transducers should also be available on site.

6. The computer system should be compatible with popular systems (e.g. IBM-PC). This will ensure that the correct hardware and software will not become obsolete.

ANNEX B

PV PLANT MONITORING : RECORD SHEET A

---

1 Name of Plant ....

2 Location ....

3 Solar Array

- cleaning date(s) ....
- change of tilt angle ....

4 Faults on System Components

- which component ....
- type of fault ....

5 Monitoring System

- down time ....
- calibration of sensors ....

6 Weather Conditions over Period

- general weather conditions ....
- extreme events (e.g. snow, hail) ....

7 Energy to Load (any unusual conditions) ....

8 Cumulative Readings of Back-up Counters

- IG ....
- JAh ....
- ELAC ....

9 Any Other Significant Event(s)

....

....

....

....

....

....

ANNEX C

PV PLANT MONITORING : RECORD SHEET B

Name of Plant .... Location ....

Date	Time	IG		JAh		ELAC		SIGNATURE
		TOTAL TO DATE (kWh/m2)	INCREMENT SINCE PREVIOUS (kWh/m2)	TOTAL TO DATE (Ah/m2)	INCREMENT SINCE PREVIOUS (Ah/m2)	TOTAL TO DATE (Ah/m2)	INCREMENT SINCE PREVIOUS (Ah/m2)	

ANNEX D

DATA SHEET FOR PV PILOT AND DEMONSTRATION PLANTS

1. Name of Plant:..
2. Location (Lat.and Longit.):.....................................
3. Rated Power of Array (kWp):......................................
   [ ] based on site measurements   [ ] based on module rated power
4. Estimated Losses in Array
   4.1 - Due to missmatch, wiring, diodes (%):......
   4.2 - Due to shading in winter (if any)(%):......
5. Orientation of Array (degrees from due North):.................
6. Tilt Angle of Array (degrees from horizontal)
   6.1 - Fixed or Initial Tilt Angle:........................
   6.2 - Range of Adjustment Possible:.......................
7. Total Area of Array = (module area)*(Nr.of modules) (m2):.......
8. Module Make and Type:............... Total Nr. of Modules:......
9. Module Dimensions (m*m):..
10. Rated Module Power (Wp):...........   NOCT (C):.............
11. Battery Size (Ah, 100 hrs):...............
12. Nominal Battery Voltage (V):..............
13. Principal Characteristics of Power Conditioning System:
    ....
    ....
    ....
14. Schematic Diagram of System (attach diagram)
15. Please attach following design information:
    15.1 - For each month
        a) Assumed average irradiation in array plane (kWh/m2)
        b) Assumed average ambient temperature (deg. C)
        c) Estimated average array output (kWh)
        d) Estimated losses in battery (kWh)
        e) Estimated losses in power conditioning equipment (kWh)
        f) Assumed load (kWh)
    15.2 - Assumed continuous power required for control and data acquisition systems
    15.3 - Number of days autonomy for which the battery storage is designed
    15.4 - Designed availability of the system (%), defined as

$$\text{Availability} = \frac{\text{Actual time in operation}}{\text{Required time in operation}} * 100\%$$

Prepared by ....   Date:.............
Address:...

LIST OF PARTICIPANTS

ANDERSEN, K.
Hasle Kommune
DK - 3790 HASLE

ARRYE, T.
Total Energie
avenue Joannes Masset 24
F - 69009 LYON

AUBERT, B.
SERI Renault
avenue du Vieil Etang 21
F - 78390 BOIS D'ARCY

AUBREE, M.
CNET
rue du Général Leclerc 38-40
F - 92131 ISSG LES MOULINEAUX

BATES, J.
Sir William Halcrow & Partners
Burderop Park
Swindon
UK - WILTS SN4 OQD

BERNARDINI, A.
Alten Consort
Via Montecarmelo 5
I - 00166 ROMA

BLAESSER, G.
CEC Joint Research Centre
I - 21020 ISPRA (VARESE)

BRICCA, E.
Pragma
Via A. d'Andrea 6
I - 00048 NETTUNO (ROMA)

CHADJIVASSILIADIS, J.
Navarinou 10
GR - 10680 ATHENS

CLAVERIE, P.
Club Alpen Français
rue la Boetie 9
F - 75008 PARIS

DICHLER, A.
Solapak
Cock Lane
UK - HIGH WYCOMBE

DROUOT, L.
Europe Energie
rue René Barthelemy 4
F - 92120 MONTROUGE

DUFRIEN, J.
CTA
Maison du Tourisme
rue République 14
F - 38000 GRENOBLE

ELIASMOLLER, H.
CEC DG XVII
rue de la Loi 200
B - 1049 BRUSSELS

FONZI, F.
Italenergie
Via della Repubblica
I - 67039 SULMONA

GILLETT, W.B.
Sir William Halcrow & Partners
Burderop Park
Swindon
UK - WILTS SN4 OQD

HULLMAN, H.
Institut 1B
Postfach 210560
D - 3000 HANNOVER

JOELS, K.
CEC DG XVII
rue de la Loi 200
B - 1049 BRUSSELS

KADELLA, H.
Siemens Solar
D - 5060 BERGISCH GLADBACH 1

KREBS, K.
CEC Joint Research Centre
I - 21020 ISPRA (VARESE)

KREMER, P.
Siemens Solar
D - 5060 BERGISCH GLADBACH 1

LAUKAMP, H.
Fraunhofer Institute, ISE
Oltmannstrasse 22
D - 7800 FREIBURG

LOWALT, H.
AEG Ag
Industriestrasse 29
D - 2000 WEDEL (HOLSTEIN)

MAASS, K.
AEG Ag
Industriestrasse 29
D - 2000 WEDEL (HOLSTEIN)

MACCIO, C.
CESEN
Via Pieragestini 50
I - GENOVA

MCCARTHY, S.
National Microelectronics
Research Centre
University College
IRL - CORK

ORSONI, M.
Pragma
Via A. d'Andrea 6
I - 00048 NETTUNO (ROMA)

PLATER, Y.
Telediffusion de France
rue de la Vanne 21
F - 92120 MONTROUGE

PREVI, A.
ENEL
Via A. Volta
I - COLOGNO MONZESE

ROESSLER, E.
Fraunhofer Institute, ISE
Oltmannsstrasse 22
D - 7800 FREIBURG

SOLECKI, J.
APREN
Le Bouloud
F - 38410 ST. MARTIN D'URIAGE

TOUSOPOULES, D.
OTE
Stadiou Street 15
GR - 124 ATHENS

VOORRAAD, J.
Hasle Kommune
DK - 3790 HASLE

WALDNER, P.
Club Alpen Français
rue la Boetie 9
F - 75008 PARIS

WEIRICH, T.
Stadtwerke Saarbrücken
Hohenzollernstrasse 104-106
D - 6600 SAARBRUCKEN

WOLFE, P.
Solapak
Cock Lane
UK - HIGH WYCOMBE

INDEX OF AUTHORS